# Biochemistry 101

## *The Easy Way*

Biochemistry I Notes
To Accompany

# Biochemistry 101

## *The Easy Way*

The Lecture Series by:

# David R. Khan, Ph.D.

E-BookTime, LLC
Montgomery, Alabama

Biochemistry 101
*The Easy Way*

ISBN: 978-1-60862-566-6

First Edition
Published August 2014
E-BookTime, LLC
6598 Pumpkin Road
Montgomery, AL 36108
www.e-booktime.com

# Chapter 1

# *Acid, Bases, Buffers, and Amino Acids*

*Chapters 1 Summary*:
*Buffers* are solutions containing a weak acid and its conjugate base or a weak base and its conjugate acid. They can resist pH changes in biological systems (i.e. Blood). Some commonly-used buffers in research include:

TAPS ({[tris(hydroxymethyl)methyl]amino}propanesulfonic acid)
Tris (tris(hydroxymethyl)methylamine)
HEPES (4-2-hydroxyethyl-1-piperazineethanesulfonic acid
MOPS 3-(N-morpholino)propanesulfonic acid

Remember that weak acids and bases have characteristic Dissociation Constants (K). The equilibrium constant for the dissociation of an acid is called the acid dissociation constant, or Ka. If we look at the dissociation of a generic weak acid:

$$HA \rightleftharpoons H^+ + A^-$$

We can express Ka as:

$$K_a = \frac{[H^+][A^-]}{[HA]}$$

Thus, the pKa is:

$$pK_a = -\log K_a$$

Remember that higher Ka values (lower pKa values) are characteristic of strong acids, while lower Ka values (high pKa values) are characteristic of weaker acids. For example, the weak acid acetic acid has a Ka value of $1.74 \times 10^{-5}$ and a pKa of 4.76. If we now look at a titration curve of acetic acid:

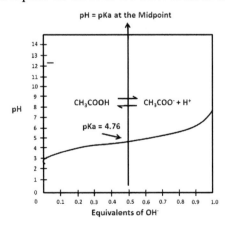

Notice that the pH is equal to the pKa at the midpoint. Another way to look at this is that you have equal concentrations of $CH_3COOH$ and $CH_3COO^-$ at pH 4.76. Also, remember from your previous Chemistry courses that you can also calculate the pH of a buffer solution if given the Ka value. So, if we were to make a buffer solution using 0.1 M acetic acid and 0.1 M sodium acetate, we could calculate the pH by determining the hydronium ion concentration:

$$Ka = \frac{[H^+][A^-]}{[HA]} = \frac{(0.1+x)(x)}{(0.1-x)}$$

$$1.74 \times 10^{-5} = x$$

$$x = [H^+]$$

$$pH = 4.76$$

A couple of things here. Notice that we made the assumption that the change (or "x") is small (less than 5%), which it is in this example (approximately 0.018%). Also, notice that we have mathematically demonstrated the observation we made with the titration curve. That is that the pH = pKa when the concentrations of both the acid and conjugate base are equal (0.1 M in both cases here). Another way to do this is with the *Henderson-Hasselbalch Equation*, which can be useful with calculations involving buffers, and as we will later see, amino acids also. If we do this, we get:

$$pH = pKa + \log \frac{[A^-]}{[HA]}$$

$$pH = 4.76 + \log \frac{[0.1]}{[0.1]}$$

$$pH = 4.76$$

Now, how is all of this related to amino acids? Well, to understand this, let's first look at the structures of the 20 common amino acids.

| Glycine | Alanine | Valine | Leucine | Isoleucine |
| Gly | Ala | Val | Leu | Ile |
| G | A | V | L | I |

| | | | | |
|---|---|---|---|---|
| Proline<br>Pro<br>P | Methionine<br>Met<br>M | Cysteine<br>Cys<br>C<br>($pK_R = 8.37$) | Serine<br>Ser<br>S | Threonine<br>Thr<br>T |
| Aspartic Acid<br>Asp<br>D<br>($pK_R = 3.90$) | Glutamic Acid<br>Glu<br>E<br>($pK_R = 4.07$) | Asparagine<br>Asn<br>N | Glutamine<br>Gln<br>Q | Lysine<br>Lys<br>K<br>($pK_R = 10.54$) |
| Arginine<br>Arg<br>R<br>($pK_R = 12.48$) | Histidine<br>His<br>H<br>($pK_R = 6.04$) | Phenylalanine<br>Phe<br>F | Tyrosine<br>Tyr<br>Y<br>($pK_R = 10.46$) | Tryptophan<br>Trp<br>W |

In order to be successful in this class, you should be familiar with the structures, names, three-letter symbols, one-letter symbols, and $pK_R$ values where they apply. You should also be familiar with the names of some of the side chains (i.e. Cysteine has a sulfhydryl group, Arginine has a guanidino group, Histidine has an imidazole group etc.). If you notice from their structures, amino acids have acid and base properties, and therefore have pK values associated with them (one for the α-carboxylic acid groups or $pK_1$ and one for the α-amino groups or $pK_2$). In addition, some of the amino acids have pK values for the side chains (or the "R" groups), and therefore are also assigned a pK value ($pK_R$). Now, back to the original question, what does our previous discussion on buffers and the Henderson-Hasselbalch Equation have to do with this? Well, the Henderson-Hasselbalch Equation can be used for calculations involving what the given amino acid looks like at a particular pH. The above amino acids structures appear this way (as drawn here) at physiological pH (7.2 - 7.4). At this pH, the zwitterionic form of the α-amino acids exists (zwitterions are molecules that have charged groups of opposite polarity). You should also note that amino acids are chiral. If you remember from Organic Chemistry, chiral molecules rotate plane polarized light either right (and are given a *d* designation) or left (and are given a *l* designation). The problem with using the classification involving optical isomers is that it provides no interpretable indication of the absolute configuration or spatial arrangement of the chemical groups about the chiral center. Furthermore, molecules that contain multiple asymmetric centers may have an overall optical rotation that is not related to the individual asymmetric centers. Therefore, we use the Fisher Convention in which the groups about the asymmetric center is related to a particular molecule that has one chiral center, namely glyceraldehyde (D-glyceraldehyde and L- glyceraldehyde). Using this convention, we can say that all of the α-amino acids are of the L- configuration.

$$\begin{array}{cc}
\text{CHO} & \text{CHO} \\
| & | \\
\text{HO}-\text{C}-\text{H} & \text{H}-\text{C}-\text{OH} \\
| & | \\
\text{CH}_2\text{OH} & \text{CH}_2\text{OH}
\end{array}$$

L- glyceraldehyde            D- glyceraldehyde

While you are not responsible for the structures of the nonstandard amino acids, it is important to point out some aspects of certain nonstandard amino acids. For example, GABA is a neurotransmitter, Thyroxine is a thyroid hormone, Histamine is involved in allergic reactions, Dopamine is a neurotransmitter, Citrulline and Ornithine are intermediates in Urea biosynthesis, Homocysteine is an intermediate in amino acid metabolism, and Azaserine is an antibiotic.

*Chapter 1*
*Lecture Series*

Slide #1 – Introduction
*Notes*

Slide #2 – Buffers are solutions containing a weak acid and its conjugate base or a weak base and its conjugate acid. They can resist pH changes in biological systems (i.e. Blood). The _____-Hasselbalch Equation can be useful in calculations involving Buffers and Amino Acids.
*Notes*

Slide #3 – A look at some common buffers used in research. It is not necessary to memorize this list.
*Notes*

Slide #4 – Weak acids and bases have characteristic Dissociation Constants (K).
*Notes*

$$HA \quad \rightleftharpoons \quad H^+ + A^-$$

$$K_a = \underline{\hspace{3cm}}$$

$$pK_a = -\log K_a$$

Slide #5 – A list of pKa values of commonly used acids. Higher K values and lower pKa values are characteristic of _____ acids, while lower K values and high pKa values are characteristic of _____ acids. Notice that acetic acid has a pKa value of 4.76.
*Notes*

Slide #6 – Titration curves. Let's look at acetic acid (pKa = 4.76). Notice that at pH 4.76, (i.e. when the pH is equal to the pKa), you have _____ % acetic acid and _____ % acetate.
*Notes*

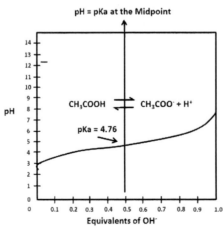

Slide #7 – How acid buffers work.
*Notes*

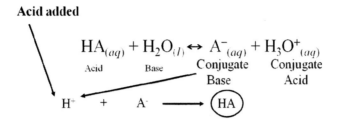

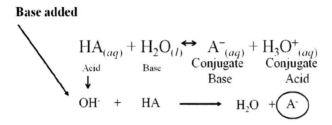

Slide #8 – What is the pH of a buffer that is 0.100 M $HC_2H_3O_2$ and 0.100 M $NaC_2H_3O_2$? Use 1.8 x $10^{-5}$ as the Ka value for this example.
*Notes*

$$HC_2H_3O_2 \longleftrightarrow H^+ + C_2H_3O_2^-$$

|  | [HA] | [A⁻] | [H⁺] |
|---|---|---|---|
| Initial |  |  |  |
| Change |  |  |  |
| Equilibrium |  |  |  |

$$K_a = \frac{[C_2H_3O_2^-][H^+]}{[HC_2H_3O_2]} = \underline{\hspace{2cm}}$$

Slide #9 – What is the pH of a buffer that is 0.100 M $HC_2H_3O_2$ and 0.100 M $NaC_2H_3O_2$? Use 1.8 x $10^{-5}$ as the Ka value for this example.
*Notes*

$$HC_2H_3O_2 \longleftrightarrow H^+ + C_2H_3O_2^-$$

|  | [HA] | [A⁻] | [H⁺] |
|---|---|---|---|
| Initial |  |  |  |
| Change |  |  |  |
| Equilibrium |  |  |  |

$$K_a = \frac{[C_2H_3O_2^-][H^+]}{[HC_2H_3O_2]} = \frac{(0.1 + X)(X)}{(0.1 - X)}$$

$$X = \underline{\hspace{2cm}}$$

Slide #10 – What is the pH of a buffer that is 0.100 M $HC_2H_3O_2$ and 0.100 M $NaC_2H_3O_2$? Use 1.8 x $10^{-5}$ as the Ka value for this example.

*Notes*

$$HC_2H_3O_2 \longleftrightarrow H^+ + C_2H_3O_2^-$$

|  | [HA] | [A⁻] | [H⁺] |
|---|---|---|---|
| Initial |  |  |  |
| Change |  |  |  |
| Equilibrium |  |  |  |

$$[H^+] = X = \underline{\hspace{2cm}}$$

$$pH = -\log [H^+] = \underline{\hspace{2cm}}$$

Slide #11 – Henderson-Hasselbalch Equation.

*Notes*

$$pH = \underline{\hspace{5cm}}$$

Slide #12 – Calculate the pKa of lactic acid, given that the concentration of free lactic acid is 0.010M and the concentration of lactate is 0.087 M (pH 4.80).

*Notes*

Slide #13 – Calculate the ratio of concentrations of acetate and acetic acid required in a buffer system of pH 5.3 (pKa=4.76).
*Notes*

.

Slide #14 – Would you use the Henderson-Hasselbalch Equation for strong acids?
*Notes*

Slide #15 – Amino Acids.
*Notes*

Slide #16 – What are the names of these amino acids?
*Notes*

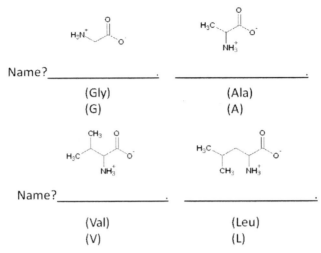

Name?_____.        _____.

       (Gly)             (Ala)
       (G)              (A)

Name?_____.        _____.

       (Val)           (Leu)
       (V)            (L)

Slide #17 – What are the names of these amino acids?
*Notes*

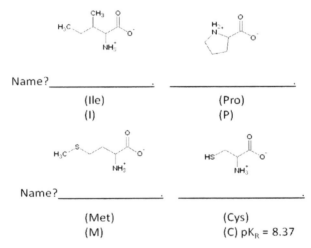

Name?_____.        _____.

       (Ile)          (Pro)
       (I)            (P)

Name?_____.        _____.

       (Met)          (Cys)
       (M)         (C) $pK_R = 8.37$

Slide #18 – What are the names of these amino acids?
*Notes*

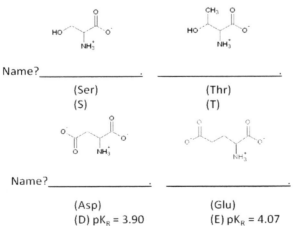

Name?_____.        _____.

       (Ser)         (Thr)
       (S)          (T)

Name?_____.        _____.

       (Asp)        (Glu)
     (D) $pK_R = 3.90$    (E) $pK_R = 4.07$

Slide #19 – What are the names of these amino acids?
*Notes*

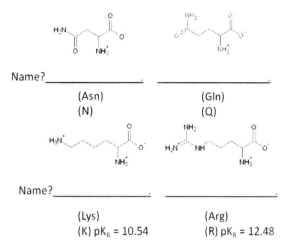

Name?_____.    _____.

(Asn)                           (Gln)
(N)                             (Q)

Name?_____.    _____.

(Lys)                           (Arg)
(K) $pK_R = 10.54$              (R) $pK_R = 12.48$

Slide #20 – What are the names of these amino acids?
*Notes*

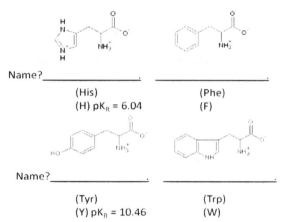

Name?_____.    _____.

(His)                           (Phe)
(H) $pK_R = 6.04$               (F)

Name?_____.    _____.

(Tyr)                           (Trp)
(Y) $pK_R = 10.46$              (W)

Slide #21 – This amino acid is _____.
*Notes*

Slide #22 – _____ are molecules that have charged groups of opposite polarity.
*Notes*

Slide #23 – Amino Acids have acid and base properties. Here is a titration curve of glycine. Notice you have a pK value for the α-carboxylic acid group ($pK_1$) and one for the α-amino group ($pK_2$).
*Notes*

Slide #24 – In a 0.1 M lysine solution, what are the individual concentrations of the protonated C-terminus form of the amino acid versus the non-protonated at pH 4 (pKa = 2.16)?
*Notes*

Slide #25 – Tetrapeptide. Amide bonds are formed between amino acids via a condensation reaction, and can undergo enzymatic hydrolysis during enzyme degradation. They are ~ 1.33 Å in length because it has double bond character, as opposed to a normal C-C bond, which is ~1.52 Å in length. What are the names of the amino acids here? Where are the amide bonds?
*Notes*

Slide #26 – Greek Lettering Scheme.
*Notes*

Slide #27 – Remember that chiral molecules rotate plane polarized light either right and are therefore dextrorotary (d ), or left and are levoratory (l ).

To be chiral, you need _____ different substituents tetrahedrally arranged about the center atom.
*Notes*

Slide #28 – The _____ Convention. The problem with using the classification involving optical isomers is that it provides no interpretable indication of the absolute configuration or spatial arrangement of the chemical groups about the chiral center. Furthermore, molecules that contain multiple asymmetric centers may have an overall optical rotation that is not related to the individual asymmetric centers. In this Convention, the groups about the asymmetric center is related to a particular molecule that has one chiral center or _____.
Can you draw the structures here?
*Notes*

Slide #29 – α-Amino Acids are of the _____ configuration.
*Notes*

Slide #30 – Cahn-Ingold-Prelog System.
*Notes*

Slide #31 – Non-standard amino acids. _____ is involved in the formation of triple helices.
*Notes*

Slide #32 – More non-standard amino acids. GABA is a neurotransmitter, Thyroxine is a hormone, _____ is involved in allergic reactions, Dopamine is a neurotransmitter, Citrulline and Ornithine are intermediates in Urea biosynthesis, homocysteine is an intermediate in amino acid metabolism, and Azaserine is an antibiotic.
*Notes*

Slide #33 – Recommended Problem Sets (from the textbook).
*Notes*

## Chapter 1
### Additional Problem Sets (test format)

1.) Which of the following statements regarding buffers is true?
   A.) A buffer solution can resist changes in pH by neutralizing acids/bases.
   B.) In a buffer made with a weak acid and its conjugate base, the weak acid can neutralize any added base.
   C.) In a buffer made with a weak acid and its conjugate base, the conjugate base can neutralize any added acid.
   D.) The pH of the buffer solution will not change that much if a strong base is added.
   E.) All of the above are true.

2.) What is the pH of a buffer solution that is 0.100 M $HC_2H_3O_2$ and 0.100 M $NaC_2H_3O_2$? The $K_a$ for $HC_2H_3O_2$ is $1.8 \times 10^{-5}$.
   A.) $1.8 \times 10^{-5}$.
   B.) 4.74.
   C.) 9.26.
   D.) 2.21.
   E.) None of the above.

3.) What is the pH of a buffer solution that is 0.200 M $HC_2H_3O_2$ and 0.200 M $NaC_2H_3O_2$? The $K_a$ for $HC_2H_3O_2$ is $1.8 \times 10^{-5}$.
   A.) $1.8 \times 10^{-5}$.
   B.) 4.74.
   C.) 9.26.
   D.) 2.21.
   E.) None of the above.

4.) What is the pH of a buffer solution that is 0.300 M $HC_2H_3O_2$ and 0.300 M $NaC_2H_3O_2$? The $K_a$ for $HC_2H_3O_2$ is $1.8 \times 10^{-5}$.
   A.) $1.8 \times 10^{-5}$.
   B.) 4.74.
   C.) 9.26.
   D.) 2.21.
   E.) None of the above.

5.) According to the Henderson-Hasselbach equation, the pH = pKa when:
   A.) [A-]/[HA] = 0.
   B.) log ([A-]/[HA]) = 1.
   C.) [A-] >> [HA].
   D.) [A-] = [HA].
   E.) None of the above.

6.) What is the ratio of monosodium citrate to citric acid (pKa = 3.09) in a 1.0 M citric acid solution with a pH = 2.09?
    A.) 10:1.
    B.) 1:1.
    C.) 1:10.
    D.) 10:11.
    E.) None of the above.

7.) Weak acids:
    A.) Are only partially ionized in aqueous solution.
    B.) Give solutions a high pH.
    C.) Do not provide hydronium ions.
    D.) Are almost insoluble in water.
    E.) None of the above.

8.) If the pH of blood is 7.4, then what is the hydronium ion concentration?
    A.) $2.5 \times 10^{-13}$ M.
    B.) $4.0 \times 10^{-2}$ M.
    C.) $4.0 \times 10^{-8}$ M.
    D.) $2.5 \times 10^{-7}$ M.
    E.) None of the above.

9.) What would the hydroxide concentration be from question #8?
    A.) $2.5 \times 10^{-13}$ M.
    B.) $4.0 \times 10^{-2}$ M.
    C.) $4.0 \times 10^{-8}$ M.
    D.) $2.5 \times 10^{-7}$ M.
    E.) None of the above.

10.) If the pH of gastric juice is 1.4, then what is the hydronium ion concentration?
    A.) $2.5 \times 10^{-13}$ M.
    B.) $4.0 \times 10^{-2}$ M.
    C.) $4.0 \times 10^{-8}$ M.
    D.) $2.5 \times 10^{-7}$ M.
    E.) None of the above.

11.) What would the hydroxide concentration be from question #10?
    A.) $2.5 \times 10^{-13}$ M.
    B.) $4.0 \times 10^{-2}$ M.
    C.) $4.0 \times 10^{-8}$ M.
    D.) $2.5 \times 10^{-7}$ M.
    E.) None of the above.

12.) What is the acetate to acetic acid (pKa = 4.76) ratio of a buffer system at pH 5.3?
  A.) 2.34.
  B.) 3.47.
  C.) 0.288.
  D.) 10.53.
  E.) None of the above.

13.) "G" is the one letter designation for which amino acid?
  A.) Glycine.
  B.) Glutamic Acid.
  C.) Glutamine.
  D.) Cysteine.
  E.) None of the above.

14.) "C" is the one letter designation for which amino acid?
  A.) Glycine.
  B.) Glutamic Acid.
  C.) Glutamine.
  D.) Cysteine.
  E.) None of the above.

15.) "E" is the one letter designation for which amino acid?
  A.) Glycine.
  B.) Glutamic Acid.
  C.) Glutamine.
  D.) Cysteine.
  E.) None of the above.

16.) "Q" is the one letter designation for which amino acid?
  A.) Glycine.
  B.) Glutamic Acid.
  C.) Glutamine.
  D.) Cysteine.
  E.) None of the above.

17.) "T" is the one letter designation for which amino acid?
  A.) Tyrosine.
  B.) Tryptophan.
  C.) Threonine.
  D.) Isoleucine.
  E.) None of the above.

18.) "W" is the one letter designation for which amino acid?
    A.) Tyrosine.
    B.) Tryptophan.
    C.) Threonine.
    D.) Isoleucine.
    E.) None of the above.

19.) "I" is the one letter designation for which amino acid?
    A.) Tyrosine.
    B.) Tryptophan.
    C.) Threonine.
    D.) Isoleucine.
    E.) None of the above.

20.) "Y" is the one letter designation for which amino acid?
    A.) Tyrosine.
    B.) Tryptophan.
    C.) Threonine.
    D.) Isoleucine.
    E.) None of the above.

21.) "Thr" is the three letter designation for which amino acid?
    A.) Tyrosine.
    B.) Tryptophan.
    C.) Threonine.
    D.) Isoleucine.
    E.) None of the above.

22.) "Trp" is the three letter designation for which amino acid?
    A.) Tyrosine.
    B.) Tryptophan.
    C.) Threonine.
    D.) Isoleucine.
    E.) None of the above.

23.) "Ile" is the three letter designation for which amino acid?
    A.) Tyrosine.
    B.) Tryptophan.
    C.) Threonine.
    D.) Isoleucine.
    E.) None of the above.

24.) Which of the following amino acids has a guanidino group?
   A.) Aspartic Acid.
   B.) Histidine.
   C.) Arginine.
   D.) Cysteine.
   E.) None of the above.

25.) Which of the following amino acids has an imidazole group?
   A.) Aspartic Acid.
   B.) Histidine.
   C.) Arginine.
   D.) Cysteine.
   E.) None of the above.

26.) Which of the following amino acids has a sulfhydryl group?
   A.) Aspartic Acid.
   B.) Histidine.
   C.) Arginine.
   D.) Cysteine.
   E.) None of the above.

27.) Which of the following is NOT a "standard" amino acid, and is an intermediate in urea biosynthesis?
   A.) Histamine.
   B.) Histidine.
   C.) Citrulline.
   D.) Cysteine.
   E.) None of the above.

28.) Which of the following is NOT a "standard" amino acid, and is involved in allergic reactions?
   A.) Histamine.
   B.) Histidine.
   C.) Citrulline.
   D.) Cysteine.
   E.) None of the above.

29.) Which of the following is NOT a "standard" amino acid, and is an antibiotic?
   A.) Phenylalanine.
   B.) Histidine.
   C.) Azaserine.
   D.) Cysteine.
   E.) None of the above.

30.) Be sure that you can draw all 20 of the "standard" amino acids on the next 2 pages.

**Glycine**

**Alanine**

**Valine**

**Leucine**

**Isoleucine**

**Proline**

**Methionine**

**Cysteine**

**Serine**

**Threonine**

**Aspartic Acid**

**Glutamic Acid**

**Asparagine**

**Glutamine**

**Lysine**

**Arginine**

**Histidine**

**Phenylalanine**

**Tyrosine**

**Tryptophan**

# Chapter 1
## Answers to Additional Problem Sets (test format)

1.) E
2.) B
3.) B
4.) B
5.) D
6.) A
7.) A
8.) C
9.) D
10.) B
11.) A
12.) B
13.) A
14.) D
15.) B
16.) C
17.) C
18.) B
19.) D
20.) A
21.) C
22.) B
23.) D
24.) C
25.) B
26.) D
27.) C
28.) A
29.) C
30.) Check your structures against those in the textbook.

# Chapter 2

# *Nucleic Acids, Gene Expression, and Recombinant DNA Technology*

*Chapter 2 Summary*:

*Nucleotides* consists of either a ribose or a 2-prime-deoxyribose residue. The C-prime atom forms a glycosidic bond with a nitrogenous base, and the 3-prime or 5-prime position is esterfied to a phosphate group. Note that the nucleotide phosphate group is doubly ionized at physiological pH's, which makes them moderately strong acids. *Ribonucleotides* are phosphate esters of a pentose sugar that contain a nitrogenous base at C1-prime. Note that the "prime" is used to describe the positioning on the sugar, which allows you to distinguish between the sugar and the positioning on the nitrogenous base. The pentose sugar found in RNA is D-ribose, which has a hydroxyl group at C2-prime. *Deoxyribonucleotides* are also phosphate esters of a pentose sugar that contain a nitrogenous base at C1-prime. Again, the "prime" is used to describe the positioning on the sugar and not the positioning on the nitrogenous base. The pentose sugar found in DNA is 2-prime-deoxy-D-ribose, which is similar to the D-ribose sugar found in RNA minus the hydroxyl group at C2-prime. Some nitrogenous bases found in nucleotides are derivatives of *Purines*, which consist of a pyrimidine ring fused to an imidazole ring. They are planar, aromatic, and heterocyclic molecules. They form a glycosidic bond in the β configuration (this configuration will be explained in more detail later when we talk about sugars) with the C1-prime of the sugar. The various positions on the structure are numbered from 1-9, with nitrogen atoms at positions 1, 3, 7, and 9. Other nitrogenous bases found in nucleotides are derivatives of *Pyrimidines*, which obviously consists of a pyrimidine ring. They are also planar, aromatic, and heterocyclic molecules. They form a glycosidic bond in the β configuration with the C1-prime of the sugar, and the various positions on the six member ring are numbered from 1-6, with nitrogen atoms at positions 1, and 3. The major Purine components of nucleic acids are *Adenine* and *Guanine*. They are present in both DNA and in RNA, and the glycosidic bond they form to ribose is via the N-9 atom. The major Pyrimidine components of nucleic acids are *Cytosine, Thymine,* and *Uracil*. Thymine occurs mainly in DNA, while Uracil is found in RNA. The glycosidic bond they form to the ribose sugar is via the N1 atoms. The structure of nucleic acids was elucidated in the 1950s by Phoebus Levine and Alexander Todd. What they found was that generally they are linear polymers of nucleotides, in which the phosphate groups bridge the 3-prime and 5-prime positions of successive sugar residues. They have directionality. At the 3-prime end, the C3-prime atom is not linked to a neighboring nucleotide, while at the 5-prime end the C5-prime atom is not linked to a neighboring nucleotide (see figure on the next page).

*Chargaff's Rule* was first described in the late 1940s by Erwin Chargaff. He determined that there are equal numbers of Adenine and Thymine residues in DNA, and that the same is true for Guanine and Cytosine. The structural basis for Chargaff's Rules is that in a double-stranded DNA molecule, Guanine is always hydrogen bonded to Cytosine (i.e. forms a base pair), and Adenine always forms a base pair with Thymine. B-DNA is the native, biologically functional form of DNA. It is a right-handed double helix composed of two antiparallel polynucleotide chains that are not identical in either base sequence or composition, rather they are complementary. Which is to say that wherever adenine appears in one chain, thymine is found in the other. The same is true with guanine and cytosine. The ideal B-DNA helix has approximately 10 base pairs per helical turn, a diameter of 20 angstroms, and a rise of 3.4 angstroms per base pair.

*Notable Experiments*:

*Ralph Brinster* microinjected DNA carrying the gene for rat growth hormone into the nuclei of fertilized mouse eggs, which were then implanted into the uteri of foster mothers. This was done in 1982, and resulted in transgenic mice (mice that carries a foreign gene that has been deliberately inserted into its genome).

*Hershey-Chase experiment*. In this experiment, bacteriophage was grown up containing radioactive isotopes. Sulfur-35 was used to label the protein capsid, while Phosphorus 32 was used to label the DNA. When this was added to E. coli and allowed to incubate for the appropriate time frame, it was determined that only the phage ghosts contained the Sulfur-35, while the bacteria itself contained most of the Phosphorus 32. Bacteriophage is any one of a number of viruses that can infect bacteria. They are among the most common organisms on Earth. The size of the nucleic acid that they carry varies depending upon the phage. The simplest phages only have enough nucleic acid to code for 3-5 average size gene products, while more complex phages may code for over 100 gene products.

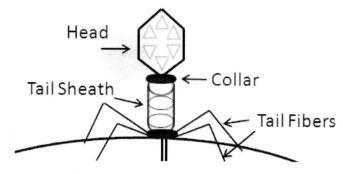

**Bacteriophage**

Another important aspect of DNA is that it can be denatured by heating for example, and can completely be renatured upon cooling, which is as process known as annealing. This will be particularly important when we talk about PCR. Notice that the shape of the absorbance curve does not change, but rather is shifted. Which is to say that it absorbs more light in the denatured form at all of the wavelengths shown here.

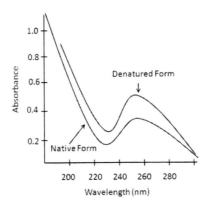

Three important concepts here are transcription, translation, and replication. During transcription, the enzyme that synthesizes RNA from the template DNA strand is RNA polymerase. This is done using nucleoside triphosphates (adenosine triphosphate (ATP), cytidine triphosphate (CTP), guanosine triphosphate (GTP), and uridine triphosphate (UTP)). During this process pyrophosphate is released. It is important to note that this is done in the 5' to 3' direction. Eukaryotic RNA undergoes post-transcriptional modifications to include gene splicing. In gene splicing, introns (intervening sequences) are precisely removed, and the remaining exons (expressed sequences) are rejoined to form mature mRNA. Alternative splicing can occur as exons of the RNA produced by transcription of a gene can be reconnected in multiple ways during RNA splicing. The resultant different mRNAs may be translated into different protein isoforms, thus a single gene may code for multiple proteins. The mRNA can then direct the ribosomal synthesis of polypeptides. Ribosomes faciliate the binding of the mRNA's codons to the anticodons of tRNAs. The tRNAs carry their corresponding amino acid (aminoacyl-tRNA). Ribosomes then catalyze the formation of peptide bonds between successive amino acids. Notice the polypeptide chain grows from the N-terminus to the C-terminus. We can use the "standard" genetic code to determine the amino acid sequence from the mRNA sequence. Obviously, the correct reading frame is important, thus the AUG sequence is used as the initiating codon to set the correct reading frame. DNA replication is similar to RNA synthesis. However, here it is synthesized from deoxynucleoside triphosphates (dNTPs) by DNA polymerase. DNA polymerase is an enzyme that can only extend existing polynucleotides bound to the DNA template, therefore primers are needed. DNA polymerase extends DNA in its 5' to 3' direction. An important technique to point out here is the polymerase chain reaction (PCR). The invention of the polymerase chain reaction in 1983 is generally credited to Kary Mullis. One of the early problems with PCR was to find a suitable DNA polymerase that could withstand the high temperatures of >90°C. This was essentially solved by using Taq polymerase (which naturally occurs in hot environments). What makes this technique so powerful is the fact that PCR can be used in order to amplify DNA to detectable levels. There are three major steps in this process. The first step is the denaturation step which involves separation of the strands of the DNA template by disrupting the hydrogen bonds between complementary bases, thereby yielding single strands of DNA. The second step is the annealing step in which the temperature is lowered to allow for the annealing of the primers to the single-stranded

DNA template. The last step is the elongation step in which DNA polymerase synthesizes a new strand of DNA that is complementary to the DNA template. The basic items needed to do this are the template DNA, two primers that are complementary to the 3' ends of each strand of DNA, *Taq polymerase* or another DNA polymerase with a temperature optimum at around 70°C, and deoxynucleoside triphosphates (dNTPs).

The last part of this chapter deals with plasmids, restriction endonucleases, and chimeric DNA. *Plasmids* are circular, double-stranded unit of DNA that replicates within a cell independently of the chromosomal DNA (autonomous replication). They contain anywhere from 1 to 200 kb pairs. Plasmids are most often found in bacteria. They are used as cloning vectors in recombinant DNA research to transfer genes between cells. They contain restriction endonuclease sites into which DNA may be inserted. *Restriction endonucleases* are enzymes that cuts double-stranded or single-stranded DNA at specific recognition nucleotide sequences known as restriction sites. Most restriction sites possess twofold rotational symmetry. These sequences are palindromes (reads the same forward and backwards). Restriction fragments can have "sticky" or "blunt" ends. Lastly, *Chimeric DNA* results when recombinant DNA (a form of DNA that does not exist naturally) is further altered or changed to host additional strands of DNA.

## Chapter 2
## *Lecture Series*

Slide #1 – Introduction
*Notes*

Slide #2 – Nucleotides contain either a _____ or a _____. The C1' atom of the sugar forms a glycosidic bond with a nitrogenous base. The nitrogenous bases found in nucleotides can be either a purine or a pyrimidine. The 3' or 5' position on the sugar is esterfied to a phosphate group, which is doubly ionized at physiological pH's.
*Notes*

Slide #3 – Ribonucleotides are phosphate esters of a pentose sugar that contain a nitrogenous base at C1'. The "prime" is used to describe the positioning on the sugar, which distinguishes it from the positioning on the _____ base.

The pentose sugar found in RNA is D-ribose, which contains a hydroxyl group at C2'. Can you draw the basic structure here?
*Notes*

**Basic Structure of a Ribonucleotide**

Slide #4 – Deoxyribonucleotides are phosphate esters of a pentose sugar that contain a nitrogenous base at C1'. Again, the "prime" is used to describe the positioning on the sugar, which distinguishes it from the positioning on the nitrogenous base. The pentose sugar found in DNA is 2'-deoxy-D-ribose, which is similar to the D-ribose sugar found in RNA, with a missing hydroxyl group at C2'. Can you draw the basic structure here?

*Notes*

**Basic Structure of a Deoxyribonucleotide**

Slide #5 – Ok, let's now focus on the nitrogenous bases. Some nitrogenous bases found in nucleotides are derivatives of Purines. They consist of a pyrimidine ring fused to an _____ ring. They are planar, aromatic, and heterocyclic molecules. They form a glycosidic bond in the β config-uration with the C1' of the sugar. The various positions on the structure are numbered from 1-9, with nitrogen atoms at positions 1, _____, _____, and 9.

*Notes*

Slide #6 – Other nitrogenous bases found in nucleotides are derivatives of Pyrimidines. They consist of a pyrimidine ring. They are planar, aromatic, and heterocyclic molecules. They form a glycosidic bond in the β configuration with the C1' of the sugar. The various positions on the six member ring are numbered from 1-6, with nitrogen atoms at positions _____ and _____.

*Notes*

Slide #7 – Purines. The major Purine components of nucleic acids are Adenine and Guanine. They are present in both DNA and in RNA. The glycosidic bond they form to ribose is via the N9 atom. Which one is which? Where is the N9 atom?

*Notes*

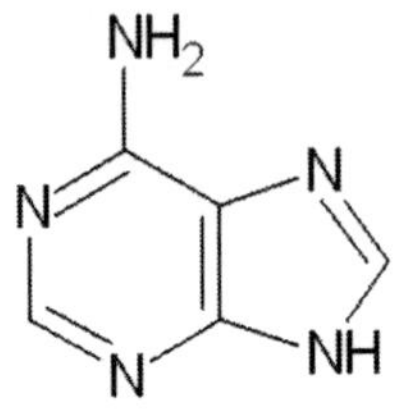

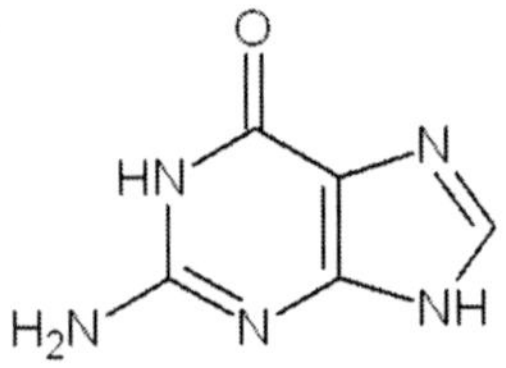

Slide #8 – The major Pyrimidine components of nucleic acids are Cytosine, Thymine, and Uracil. Thymine occurs mainly in DNA and Uracil in RNA. The glycosidic bonds they form to the ribose sugar is via the N1 atoms. Which one is which? Where is the N1 atom?

*Notes*

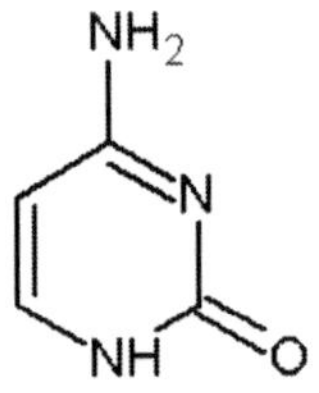

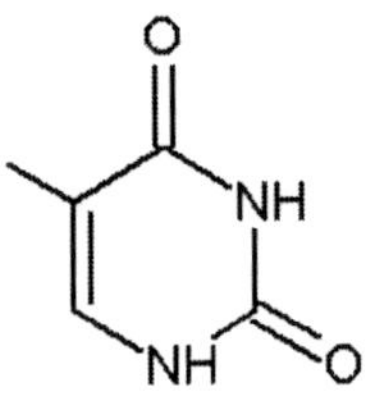

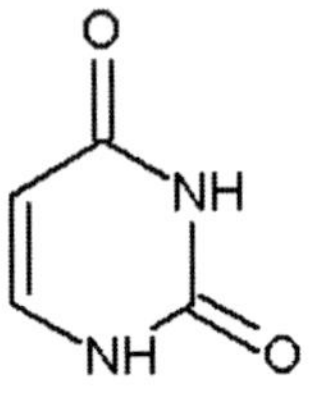

Slide #9 – Common modifications of adenine and cytosine in the DNAs of many organisms include are N6-Methyl-dA and 5-Methyl-dC.

*Notes*

Slide #10 – Structure of nucleic acids elucidated in the 1950s by Phoebus Levine and Alexander Todd. They are generally linear polymers of nucleotides. The phosphate groups of the nucleotides bridge the 3' and 5' positions of successive sugar residues. Polynucleotides have_____.
The 3' end is where the C3' atom is not linked to a neighboring nucleotide, whereas the 5' end is where the C5'atom is not linked to a neighboring nucleotide.
*Notes*

Slide #11 – Chargaff's Rules were first described in the late 1940s by Erwin Chargaff. It was determined that there are equal numbers of Adenine and Thymine residues in DNA. There are also equal numbers of Guanine and Cytosine residues in DNA. In double-stranded DNA, _____ is always hydrogen bonded to Cytosine (i.e. forms a base pair), while _____ always forms a base pair with Thymine.
*Notes*

Slide #12 – B-DNA is the _____, biologically functional form of DNA. It is a right-handed double helix composed of two antiparallel polynucleotide chains that are complementary to one another with a diameter of ~20 Å. B-DNA has a rise of ~3.4 Å/bp, and there are ~10 bp/helical turn (see picture in your book). Is this a right-handed helix?
*Notes*

Slide #13 – Base pairing. Adenine and Thymine base pairs are joined by _____ Hydrogen bonds, and Guanine and Cytosine base pairs are joined by _____ Hydrogen bonds.
*Notes*

Slide #14 – Transgenic Mice are mice that carry a foreign gene that has been deliberately inserted into its genome.
*Notes*

Slide #15 – _____ is any one of a number of viruses that infect bacteria. Bacteriophages are among the most common organisms on Earth. The size of the nucleic acid varies depending upon the phage. The simplest phages only have enough nucleic acid to code for 3-5 average size gene products, while more complex phages may code for over 100 gene products.
*Notes*

Slide #16 – Bacteriophage Structure. Can you fill in the blanks?
*Notes*

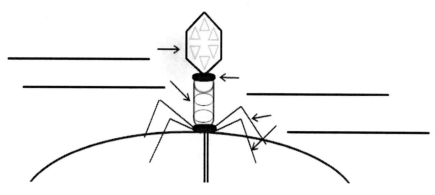

Slide #17 – In the _____ - _____ experiment, bacteriophage was grown containing radioactive isotopes (either $^{35}$S-labelled shell or $^{32}$P-labelled DNA). It was added to E. coli, and it was determined that only the phage ghosts contained the $^{35}$S, while the bacteria contained most of the $^{32}$P.
*Notes*

Slide #18 – DNA of a bacteriophage which had been osmotically lysed in distilled water. This DNA has been fattened using the _____ procedure (coating with a denatured basic protein) to ~200 Å in diameter.
*Notes*

Slide #19 – DNA can be denatured by heating, and can completely be renatured upon cooling (annealing). The shape of the absorbance curve does not change. The denatured form of DNA absorbs more light at all of the wavelengths shown here. Which one is which (i.e. which one is the native and which one is the denatured form)?
*Notes*

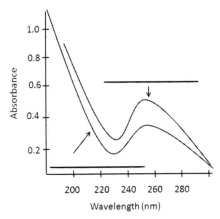

Slide #20 – DNA directs its own replication. RNA is produced via Transcription from ribonucleoside triphophates on DNA templates by RNA _____. The DNA template is read from its 3' to 5' direction and RNA is synthesized in its 5' to 3' direction. Translation of the RNA yields proteins.
*Notes*

Slide #21 – RNA Polymerase synthesizes RNA using nucleoside triphosphates (adenosine triphosphate (ATP), cytidine triphosphate (CTP), guanosine triphosphate (GTP), and uridine triphosphate (UTP)) and releases pyrophosphate.
*Notes*

Slide #22 – The following DNA molecule is transcribed from right to left. Identify the template strand.
*Notes*

5'-TC...............................CA-3'
3'-AG...............................GT-5'

Slide #23 – Eukaryotic RNA undergoes post-transcriptional modifications to include gene splicing. In gene splicing, _____ (intervening sequences) are precisely removed, and the remaining _____ (expressed sequences) are rejoined to form mature mRNA.
*Notes*

Slide #24 – Exons of the RNA produced by transcription of a gene are reconnected in multiple ways during RNA splicing. The resultant different mRNAs may be translated into different protein isoforms. Thus, a single gene may code for multiple proteins.
*Notes*

Slide #25 – A look at alternative splicing. See how these can be connected differently.
*Notes*

Slide #26 – Polypeptide Synthesis. Now that we have our mRNA, mRNAs direct the ribosomal synthesis of polypeptides. Ribosomes facilitate the binding of the mRNA's codons to the anticodons of _____. The tRNAs carry their corresponding amino acid (referred to as an aminoacyl-tRNA). _____ then catalyze the formation of peptide bonds between successive amino acids.
*Notes*

Slide #27 – Polypeptides are synthesized using the mRNA by ribosomes. Transfer RNAs (tRNAs) deliver amino acids to the ribosomes. The anticodon sequence on tRNAs are complementary to codon sequences which specify the amino acid to be added next. The polypeptide chain grows from the N-terminus to the C-terminus.
*Notes*
P-site = _____.
A-site = _____.

Slide #28 – Transfer RNA. The t-RNA molecule has a "_____" structure. The aminoacyl t-RNA has the covalently linked amino acid on the top, while the _____ is on the bottom as depicted here. This is a trinucleotide sequence that base pairs with the complementary codon on mRNA during translation.
*Notes*

Slide #29 – The "Standard" Genetic Code.
*Notes*

Slide #30 – Is the reading frame important to acquiring the correct amino acid sequence? Thus, we need an initiating codon to set the reading frame. Translation is initiated at specific _____ codons, which specifies a _____ amino acid residue.
*Notes*

Slide #31 – What is the polypeptide sequence given the following mRNA Sequence?
*Notes*

Slide #32 – DNA synthesis is similar to RNA synthesis. DNA is synthesized from deoxynucleoside triphosphates (dNTPs) by DNA polymerase. DNA polymerase is an enzyme that can only extend existing polynucleotides bound to the DNA template, therefore _____ are needed. DNA polymerase extends DNA in its 5' to 3' direction.
*Notes*

Slide #33 – The Polymerase Chain Reaction (PCR). The invention of the polymerase chain reaction in 1983 is generally credited to Kary Mullis. One of the early problems with PCR was to find a suitable DNA polymerase that could withstand the high temperatures of >90°C. This was solved by using *Taq polymerase* (which naturally occurs in hot environments).
*Notes*

Slide #34 – The Polymerase Chain Reaction (PCR) can be used in order to amplify DNA to detectable levels using three major steps. 1.) Denaturation step- Separates the strands of the DNA template by disrupting the hydrogen bonds between complementary bases, thereby yielding single strands of DNA. 2.) The _____ step-The temperature is lowered to allow for the annealing of the primers to the single-stranded DNA template. 3.) Elongation step in which DNA polymerase synthesizes a new strand of DNA that is complementary to the DNA template. *Notes*

Slide #35 – To complete PCR we need the following items:
*Notes*

    1.) _____.

    2.) Two _____ that are complementary to the 3' ends of each strand of DNA.

    3.) _____ polymerase or another DNA polymerase with a temperature optimum at around 70°C.

    4.) Deoxynucleoside triphosphates (dNTPs).

Slide #36 – _____ are circular, double-stranded unit of DNA that replicates within a cell independently of the chromosomal DNA (autonomous replication). They contain anywhere from 1 to 200 kb pairs. They are most often found in bacteria. They are used as cloning vectors in recombinant DNA research to transfer genes between cells. They contain restriction _____ sites into which DNA may be inserted.
*Notes*

Slide #37 – Restriction Endonucleases are enzymes that cuts double-stranded or single-stranded DNA at specific recognition nucleotide sequences known as restriction sites. Most restriction sites possess twofold rotational symmetry. These particular sequences are _____ (i.e. they read the same forward and backwards). Restriction fragments can have "sticky" or "blunt" ends.

*Notes*

Slide #38 – Examples of "sticky" or "blunt" ends. EcoRI recognizes GAA*TTC resulting in DNA fragments with "sticky ends", while EcoRV recognizes GA*TATC resulting in DNA fragments with "blunt ends".

*Notes*

Slide #39 – Chimeric DNA results when recombinant DNA is further altered or changed to host additional strands of DNA. A "Chimera" is a monster from Greek mythology that has a lion's head, a goat's body, and a serpent's tail.

*Notes*

Slide #40 – Recommended Problem Sets (from the textbook).
*Notes*

## Chapter 2
### *Additional Problem Sets (test format)*

1.) Which of the following statements regarding nucleotides is/are true?
    A.) They contain either a ribose or a 2'-deoxyribose residue.
    B.) The C1' atom of the sugar forms a glycosidic bond with a nitrogenous base.
    C.) Nitrogenous bases found in nucleotides can be either a purine or a pyrimidine.
    D.) There is a phosphate group on the sugar.
    E.) All of the above are true.

2.) Nucleotides contain phosphate groups bonded to the:
    A.) C3 or C5 atoms.
    B.) C3 or C3 atoms.
    C.) C5 or N9 atoms.
    D.) C3' or C5' atoms.
    E.) None of the above.

3.) According to Chargaff's Rules, which of the following is true?
    A.) There equal numbers of Adenine and Thymine residues in DNA.
    B.) There are equal numbers of Guanine and Cytosine residues in DNA.
    C.) In double-stranded DNA, Guanine is always hydrogen bonded to Cytosine.
    D.) In double-stranded DNA, Adenine always forms a base pair with Thymine.
    E.) All of the above are true.

4.) What is the approximate molecular mass of a segment of B-DNA that specifies a 10-kD protein? (Assume the average molecular mass of an amino acid is 110 D and 288 D for deoxynucleotide residues).
    A.) 157 D.
    B.) 943 D.
    C.) 157 kD.
    D.) 943 kD.
    E.) None of the above.

5.) What is the approximate molecular mass of a segment of B-DNA that specifies a 60-kD protein? (Assume the average molecular mass of an amino acid is 110 D and 288 D for deoxynucleotide residues).
    A.) 157 D.
    B.) 943 D.
    C.) 157 kD.
    D.) 943 kD.
    E.) None of the above.

6.) What is the approximate molecular mass of a segment of B-DNA that specifies a 30-kD protein? (Assume the average molecular mass of an amino acid is 110 D and 288 D for deoxynucleotide residues).
   A.) 471kD.
   B.) 943 kD.
   C.) 157 kD.
   D.) 236 kD.
   E.) None of the above.

For questions #7 - #10, consider the following depiction of a bacteriophage:

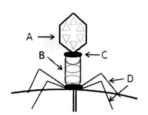

7.) The "A" would be the _____.
   A.) Tail Fibers.
   B.) Collar.
   C.) Tail Sheath.
   D.) Head.
   E.) None of the above.

8.) The "B" would be the _____.
   A.) Tail Fibers.
   B.) Collar.
   C.) Tail Sheath.
   D.) Head.
   E.) None of the above.

9.) The "C" would be the _____.
   A.) Tail Fibers.
   B.) Collar.
   C.) Tail Sheath.
   D.) Head.
   E.) None of the above.

10.) The "D" would be the _____.
   A.) Tail Fibers.
   B.) Collar.
   C.) Tail Sheath.
   D.) Head.
   E.) None of the above.

11.) Which of the following is true regarding the Hershey-Chase experiment?
    A.) It was tested in 1952 by Alfred Hershey and Martha Chase.
    B.) The phage particle contained a $^{35}$S-labeled shell.
    C.) The phage particle initially contained $^{32}$P-labeled DNA.
    D.) Following the experiment, it was determined that only the phage ghosts contained the $^{35}$S, while the bacteria contained most of the $^{32}$P.
    E.) All of the above are true.

For questions #12 - #14, consider the following spectra (UV-absorbance) of native and heat denatured DNA.

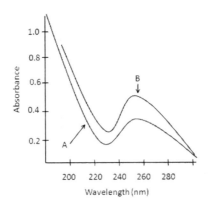

12.) "A" would be _____ DNA.
    A.) Native.
    B.) Denatured.
    C.) Naive.
    D.) Denial.
    E.) None of the above.

13.) "B" would be _____ DNA.
    A.) Native.
    B.) Denatured.
    C.) Naive.
    D.) Denial.
    E.)None of the above.

14.) Which of the following is true regarding the above spectra?
    A.) "B" would be obtained at a higher temperature than "A".
    B.) "A" would be obtained at a higher temperature than "B".
    C.) The temperatures are probably equal.
    D.) Temperature has nothing to do with these spectra.
    E.) None of the above.

15.) In order to perform PCR, which of following is necessary?
    A.) The template DNA fragment.
    B.) Primers flanking the region of interest.
    C.) dNTPs.
    D.) DNA Polymerase.
    E.) All of the above are necessary to perform PCR.

For questions #16 - #20, use the Standard Genetic Code chart from your textbook to answer the following questions.

16.) The polypeptide that would be generated given the mRNA sequence
5'- AUG UGC CAG AGC UAG-3' would be _____.
    A.) MCQS.
    B.) MDGS.
    C.) MCGS.
    D.) MGCS.
    E.) None of the above.

17.) The polypeptide that would be generated given the mRNA sequence
5'-CAU AUG UGC CAG AGC UAG-3' would be _____.
    A.) HMCQS.
    B.) MCQS.
    C.) MDGS.
    D.) MCGS.
    E.) None of the above.

18.) The polypeptide that would be generated given the mRNA sequence
5'-CAU AUG GAC GGA AGC UAA-3' would be _____.
    A.) HMCQS.
    B.) MCQS.
    C.) MDGS.
    D.) MCGS.
    E.) None of the above.

19.) The polypeptide that would be generated given the mRNA sequence
5'-CAU AUG UGG CCG CUU UAA CGA-3' would be _____.
    A.) HMYS.
    B.) MGQR.
    C.) MDGS.
    D.) MWPL.
    E.) None of the above.

20.) The polypeptide that would be generated given the mRNA sequence
5'-CAU AUG UGG CCG CUU UUU AUU UGA CGA-3'

would be _____.
A.) HMYSIF.
B.) MGQRR.
C.) MWPLFI.
D.) MWPIFL.
E.) None of the above.

21.) Small circular DNA molecules used to carry foreign DNA are called _____.
A.) Bacteriophages.
B.) Plasmids.
C.) Lipids.
D.) Carbohydrates.
E.) None of the above.

22.) The restriction fragments that result when using EcoRI are commonly referred to as having _____.
A.) Sticky ends.
B.) Blunt ends.

23.) The restriction fragments that result when using EcoRV are commonly referred to as having _____.
A.) Sticky ends.
B.) Blunt ends.

24.) As most restriction sites posses twofold rotational symmetry (i.e. they read the same forward and backwards), we commonly refer to these sequences as being _____.
A.) Plasmids.
B.) Chimeric.
C.) Palmitoyl.
D.) Palindromic.
E.) None of the above.

25.) Which of the following DNA sequences is palindromic?
A.) AGA.
B.) TTAATT.
C.) TAT.
D.) AGCT.
E.) None of the above.

26.) Which of the following DNA sequences is palindromic?
    A.) AGT.
    B.) AGGA.
    C.) AAGCTT.
    D.) AGCA.
    E.) None of the above.

27.) Which of the following DNA sequences is palindromic?
    A.) CAT.
    B.) CAC.
    C.) TATA.
    D.) AGCA.
    E.) None of the above.

28.) Which of the following DNA sequences is palindromic?
    A.) TAT.
    B.) GAG.
    C.) CGCG.
    D.) TGCT.
    E.) None of the above.

29.) Which of the following DNA sequences is palindromic?
    A.) ATAT.
    B.) CTCT.
    C.) TGCT.
    D.) GCT.
    E.) None of the above.

30.) On the next page, be sure that you can draw both Purines, and all three of the Pyrimidines.

**Purines**

Adenine

Guanine

**Pyrimidines**

Cytosine

Thymine

Uracil

## Chapter 2
### Answers to Additional Problem Sets (test format)

1.) E
2.) D
3.) E
4.) C
5.) D
6.) A
7.) D
8.) C
9.) B
10.) A
11.) E
12.) A
13.) B
14.) A
15.) E
16.) A
17.) B
18.) C
19.) D
20.) C
21.) B
22.) A
23.) B
24.) D
25.) D
26.) C
27.) C
28.) C
29.) A
30.) Check your structures against those in the textbook.

# Chapter 3

## *Techniques of Protein and Nucleic Acid Purification*

*Chapter 3 Summary*:

The first step in solubilzation is getting the protein out of the cell, or cell lysis. This can be achieved through *Osmotic lysis* using a hypotonic solution, *lysozymes* which are enzymes that can degrade bacterial cell walls, *detergents* such as Triton-X 100, and *organic solvents* (acetone or tolulene). Mechanical force can also be used. This includes *Grinding*, the use of a *high-speed blender*, *homogenizer* which uses a piston to crush cells, a *French Press* which uses high pressure to sheer cells, or a *Sonicator* which breaks cells open with Ultrasonic vibrations. Once removed from the natural environment, proteins are exposed to agents such as proteases, which can potentially irreversibly damage the protein that you may be looking for. Proteases are enzymes that catalyze the hydrolytic cleavage of peptide bonds. Therefore, what is commonly used is a Buffer (i.e. Radioimmunoprecipitation assay buffer-RIPA buffer) that contains protease inhibitors. This buffer allows for the disruption of cell membranes because it contains a various components such as Triton-X 100. Protease inhibitors are commercially available. Due to the fact that proteins are generally present in very small quantities, assays of proteins must be highly sensitive to their presence, and very specific for the protein of interest. They must also be convenient to use as it will have to be done repeatedly. *Immunochemical techniques* are highly sensitive techniques which can readily detect small amounts of specific proteins. They involve the use of antibodies, which are proteins produced by an animal's immune system in response to an antigen (or a foreign substance placed in the body), and they are also commercially available. An ELISA ("enzyme-linked immunosorbent assay") utilizes an immobilized antibody against the protein of interest on a solid support (i.e. polystyrene). Cell lysates are then added and the unbound proteins are washed away. Then a secondary antibody which is covalently linked to an easily assayed enzyme is added. The amount of protein present in the sample is then determined by enzymatic activity following substrate addition. This is just one of several techniques that are discussed in this chapter. However, before we get to some others, let's discuss protein solubilites for a minute. When proteins fold, they tend to bury the hydrophobic residues in their core, although some do remain at the surface. The hydrophilic surface is then mostly exposed to the solvent. We can potentially increase the solubility of a protein using a technique known as "*salting in*". At low salt concentrations, the solubility of proteins increases with the addition of salt. This is because the ions from the salt associate with the surface of the protein and shield the protein from water molecules. Thereby less water molecules are required to interact with the protein surface, which has the effect of increasing the "free" water, which results in an increase in protein solubility. At high salt concentrations, the opposite happens. During "*salting out*", all of the binding sites on the protein surface for the salt ions have become occupied, and the salt ions now begin to interact with the solvent (i.e. water). This has the effect of decreasing the concentration of "free" water molecules as they are now being used to solvate the excess salt ions. Protein molecules therefore begin to precipitate out of the solution.

*pH Effects.* Proteins have many ionizable groups, which results in a net charge. However, at the *Isoelectric Point*, which is specific for each protein, it carries no net charge. Proteins above the isoelectric point are predominantly negatively charged. Due to the electrostatic repulsion, there is minimal protein aggregation and the protein is soluble. Proteins below the isoelectric point are predominantly positively charged, and again due to the electrostatic repulsion, there is minimal protein aggregation and the protein is still soluble. However, proteins at their isoelectric point have no net charged, and protein aggregation decreases solubility and allows them to precipitate out of solution. This is known as *Isoelectric Precipitation*.

*Chromatography.* This is a term that involves various laboratory techniques for the separation of mixtures, which involves both a mobile phase and a stationary phase. In column chromatography, the mobile phase is the mixture of substances to be fractionated or separated, and the stationary phase is the porous solid matrix in the column. The solution to be injected is usually called a sample, and the individually separated components are called analytes. Here we will discuss three major types of column chromatography techniques (HPLC discussed later). Specifically, Ion exchange, Gel Filtration, and Affinity chromatography. First, Ion exchange chromatography involves ions electrostatically bound to the column which are replaced by ions in a solution. This technique allows for the separation of ions and polar molecules based on their charge properties. Cation exchange chromatography utilizes a negatively charged resin to bind positively charged ions. Anion exchange chromatography utilizes a positively charged column to bind negatively charged ions. Proteins that have a weak electrostatic attraction to the column come off quickly, while proteins with a high affinity for the column are gradually "pulled off" the column by increasing the salt concentration.

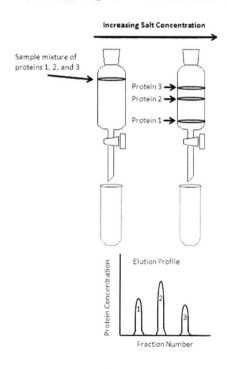

Two useful ion exchangers that you should be familiar with are DEAE-cellulose (which is primarily used to separate acidic and neutral proteins) and CM-cellulose (which is primarily used to separate basic and neutral proteins). In Gel Filtration (aka size exclusion) chromatography, molecules are separated according to size. Here, the stationary phase contains a hydrated spongelike material called gel beads, which are commonly either dextran-based or agarose-based. The gel beads have various pore sizes, which are selected based on the sizes of the proteins being separated. Here, smaller molecules get caught in the stationary phase and come off the column a lot slower than larger ones which will come off quickly. We can calculate the relative elution volume ($V_e/V_o$), where $V_e$ is the volume required to elute the sample, while $V_o$ is the volume surrounding the beads. Thus, it follows that larger proteins will have low Relative Elution volumes, while smaller proteins have larger Relative Elution volumes as they are caught in the stationary phase and take longer to come off the column, which requires more running buffer. This technique can be used to desalt a protein. Two very commonly-used gels are Sephadex (dextran-based) and Sepharose (agarose-based). Another technique that can also be used to separate based on size is dialysis (which is not a column chromatographic technique). The final type of column chromatography that we are going to discuss is Affinity chromatography, which is based on the proteins' affinity for an immobilized ligand. The binding between the protein and the stationary phase is based on tight noncovalent interactions. The stationary phase contains a ligand that specifically binds the protein of interest, which is covalently bound to a solid resin matrix (usually agarose). One of the reasons that agarose is commonly-used is due to the fact that it has numerous free hydroxyl groups that can be "activated" in order to bind the ligand. One issue here can be protein/ligand recognition due to the steric interference attributed to the big bulky agarose group. This can be resolved with commercially available "epoxy-activated" agarose resins (with for example a $C_{12}$ spacer arm). The epoxy group is free to react with a number of nucleophilic groups present on various ligands. To remove the desired protein from the column following all of the impurities that come off first, one of two things can generally be done. One, you can add a new elution solution to the column that contains a compound with a higher affinity for the protein-binding site than the bound ligand. Or, you can alter the solution conditions to destabilize the protein/ligand complex (i.e. changes in pH, ionic strength, and/or temperature).

*Other Techniques*
Paper chromatography. In this procedure, the paper is the stationary phase and the organic solvent (or a mixture of solvents) is used as the mobile phase. You spot a few drops of the sample solution on the paper (~2 cm from the bottom). You then place the paper in a container that has organic solvents. The proteins move up the paper via capillary action. Smaller proteins move further up the paper than larger ones. The migration rate can be expressed as follows:

$$R_f = \frac{\text{The distance traveled by a substance}}{\text{The distance traveled by the solvent}}$$

This equation allows you to calculate the distance your substance travelled up the paper while accounting for how far the solvent front moved.

*Electrophoresis.* This technique involves the migration of ions in an electric field, and was first reported in 1937 by Swedish biochemist Arne Tiselius, who won the Nobel Prize in Chemistry in 1948. Here, the dispersed particles have an electric surface charge. Thus, when an external electric field is applied, the particles begin to migrate. With respect to paper electrophoresis, the sample is spotted in the middle of a piece of filter paper whose ends are placed in buffer solutions that contain oppositely charged electrodes. The application of a current causes ions to migrate toward the electrode of opposite polarity. In gel electrophoresis, the gel retards the mobility of larger molecules, so smaller molecules travel further, and the most common gels used in this technique are made of polyacrylamide (PAGE) and agarose. Gel electrophoresis produces what is known as an electrophoretogram. For the separation of proteins, sodium dodecyl sulfate (SDS) is commonly-used. SDS is a detergent that binds proteins and causes them to assume a rodlike shape. In addition, the large charge that SDS imparts on proteins masks their individual charges. Thus, protein separation is based on differences in the molecular masses proteins, and not on differences in shape or charge. For a western blot, a gel is run resulting in an electrophoretogram. This is then blotted on a sheet of nitrocellulose paper membrane (commonly referred to as a transfer). The membrane is then blocked to prevent nonspecific binding, and is then incubated with the primary antibody which recognizes the protein of interest. Following several rinses, the membrane is then incubated with a secondary antibody that has been covalently linked to an enzyme (i.e. horseradish peroxidase (HRP)). The membrane is then again rinsed several times and the substrate is added to view the protein band of interest. In 2D gel electrophoresis, the proteins are first electrophoresed through a solution having a pH gradient in which the pH smoothly increases. The proteins migrate to the position in the pH gradient that corresponds to their isoelectric point (known as Isolelectric Focusing). Proteins are then further separated by SDS-PAGE. Electrophoresis can also be used for nucleic acid separation. However, due to their larger size, agarose is generally used rather than polyacrylamide. Also, bands are generally detected using ethidium (most common), acridine orange, or proflavin.

# Chapter 3
## Lecture Series

Slide #1 – Introduction
*Notes*

Slide #2 – The first step in solubilzation is getting the protein out of the cell via cell lysis. This can be achieved through osmotic lysis using a hypotonic solution, lysozymes (enzymes that can degrade bacterial cell walls), detergents (i.e. Triton-X 100), and organic solvents. Mechanical force can also be used. This includes Grinding, the use of a high-speed blender, homogenizer, a French Press, and/or a Sonicator.
*Notes*

Slide #3 – Once removed from the natural environment, proteins are exposed to various agents than can irreversibly damage the protein. _____ are enzymes that catalyze the hydrolytic cleavage of peptide bonds. For the disruption of cell membranes, buffers can be used (i.e. Radioimmunoprecipitation assay buffer-RIPA buffer). Protease inhibitors (commercially available) can be added to prevent damage to the protein.
*Notes*

Slide #4 – A look at the process.
*Notes*

Slide #5 – Assay of proteins. Due to the fact that proteins are generally present in very small quantities, assays of proteins must be highly sensitive to its presence, and very specific for the protein of interest. They must also be convenient to use, because it may have to be done repeatedly at every stage of the purification process.
*Notes*

Slide #6 – _____ techniques are highly sensitive techniques which can readily detect small amounts of specific proteins. They use antibodies, which are proteins produced by an animal's immune system in response to an antigen. There are antibodies that are commercially available for many proteins of different species.
*Notes*

Slide #7 _____ stands for enzyme-linked immunosorbent assay. Here, you have an immobilized antibody against the protein of interest on a solid support (i.e. polystyrene). Cell lysates are then added and unbound protein washed away. A secondary antibody which is covalently linked to an easily assayed enzyme is then added. The amount of protein present in the sample is then determined by enzymatic activity following substrate addition. **Something to think about:** Why is it necessary to have two antibodies rather than one?

*Notes*

Slide #8 – When proteins fold, they tend to bury _____ residues in their core, although some do remain at the surface. The _____ surface is then mostly exposed to the solvent. Therefore, the surface contains charged, uncharged and hydrophobic regions.

*Notes*

Slide #9 – "_____ in". At low salt concentrations, the solubility of proteins increases with the addition of salt. This is because the ions from the salt associate with the surface of the protein and shield the protein from water molecules. Thereby less water molecules are required to interact with the protein surface, and the concentration of "_____" water is increased resulting in an increase in protein solubility.

*Notes*

Slide #10 – "_____ out". At high salt concentrations, the solubility of proteins decreases and can precipitate out of solution. This is because all of the binding sites on the protein surface for the salt ions have become occupied, and the salt ions begin to interact with the solvent (i.e. water). This has the effect of decreasing the concentration of "_____" water molecules as they now are being used to solvate the excess salt ions. Protein molecules therefore begin to precipitate out of the solution.
*Notes*

Slide #11 – Proteins have many ionizable groups, which results in a net charge. However, at the _____ Point, they carry no net charge. Proteins above the isoelectric point are predominantly _____ charged. Due to the electrostatic repulsion, there is minimal protein aggregation and the protein is soluble. Proteins below the isoelectric point are predominantly _____ charged, and again due to the electrostatic repulsion, there is no protein aggregation and the protein is still soluble. However, proteins at their isoelectric point have no net charge, and protein aggregation decreases solubility and allows them to precipitate out of solution. This is known as Isoelectric _____.
*Notes*

Slide #12 – Solubility of _____ which has an isoelectric point of _____.
*Notes*

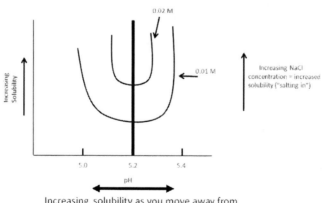

Increasing solubility as you move away from the isoelectric point in either direction

Slide #13 – _____. This technique involves a mobile phase which is a mixture of substances to be fractionated or separated, and a stationary phase (which is the porous solid matrix in the column for column chromatography). Here we have Ion exchange, in which ions electrostatically bound to the column are replaced by ions in solution. Gel Filtration in which molecules are separated according to size and Affinity chromatography which is based on the proteins affinity for an immobilized ligand.
*Notes*

Slide #14 – Ion Exchange Chromatography (IEC) allows for the separation of ions and polar molecules based on their charge properties. Cation exchange chromatography utilizes a _____ charged resin to bind positively charged ions, while anion exchange chromatography utilizes a _____ charged column to bind negatively charged ions. The solution to be injected is usually called a sample, and the individually separated components are called analytes.
*Notes*

Slide #15 –
*Notes*

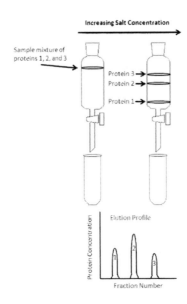

Slide #16 – Useful Ion Exchangers. Which ones should you know?
*Notes*

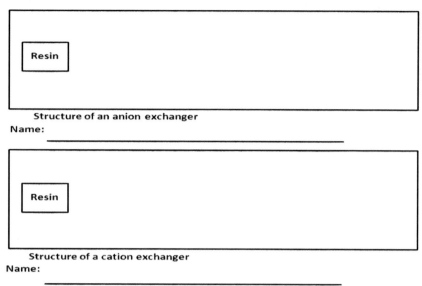

Structure of an anion exchanger
Name: _____

Structure of a cation exchanger
Name: _____

Slide #17 – _____ Filtration is also called _____ Exclusion and Molecular Sieve chromatography. Molecules are separated according to size and shape. The stationary phase contains hydrated spongelike material called gel beads, which is commonly either dextran-based or agarose-based. The gel beads have various pore sizes, which are selected based on the sizes of the proteins being separated. They commonly range from 0.05 kD to 150,000 kD.
*Notes*

Slide #18 –
*Notes*

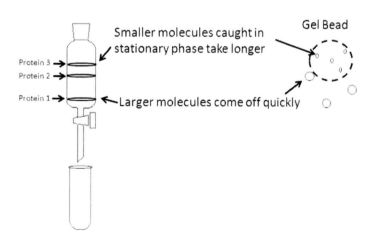

63

Slide #19 – Elution Profiles. _____ proteins have a very high relative elution volume, while _____ proteins have a very low relative elution volume.

*Notes*

$V_e$ = _____.

$V_o$ = _____.

$V_e$ / $V_o$ = _____.

Slide #20 – Commonly-used gel filtration materials. You should know that if you have a _____ column, then that is size exclusion.

*Notes*

Slide #21 – _____ is another technique used to separate based on size (i.e. separate small and large molecules).

*Notes*

Slide #22 – _____ chromatography. Separation is based on protein affinity for an immobilized ligand. The affinity for the stationary phase is based on tight _____ interactions. The stationary phase contains a ligand that specifically binds the protein of interest. The ligand is covalently bound to solid resin matrix (usually agarose). Agarose has numerous free hydroxyl groups that can be "activated" in order to bind the ligand.
*Notes*

Slide #23 – The protein's affinity for the matrix-anchored ligand determines how quickly it comes off the column. Agarose is used as the stationary phase, which can be activated with _____ prior to ligand addition.

*Notes*

Slide #24 – One of the problems with this technique can be Protein-ligand recognition due to steric interference attributed to the big bulky agarose group. This can be resolved with commercially available "_____ - _____" agarose resins, which can be purchased with a C12 spacer arm. The epoxy group is then free to react with a number of nucleo-philic groups present on various ligands.
*Notes*

Slide #25 – The epoxy group reacts with many nucleophilic groups on ligands.
*Notes*

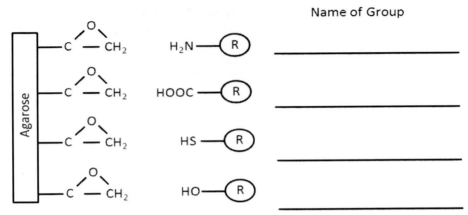

Name of Group

_____

_____

_____

_____

Slide #26 – How does the protein come off the column?

1.) One, you can add a new elution solution to the column that contains a compound with a _____ for the protein-binding site than the bound ligand.

2.) Or, you can alter the solution conditions to destabilize the protein-ligand complex (changes in _____, _____, and/or _____).

*Notes*

Slide #27 – Purification of Staphylococcal Nuclease. Protein detection monitored via absorbance at _____ nm. Notice that the first peak is everything you don't want (unwanted protein with no nuclease activity). The protein you want (Staphylococcal Nuclease) is bound to the column and does not come off until you drop the pH. Notice at this point both the absorbance and nuclease activity goes up.

*Notes*

Slide #28 – Paper chromatography. In this procedure, the _____ is the stationary phase and the organic solvent (or a mixture of solvents) is used as the mobile phase. You spot a few drops of the sample solution on the paper (~2 cm from the bottom). You then place the paper in a glass jar containing the organic solvents, and the proteins move up the paper via _____ action (which is the flow of liquids through porous media). _____ proteins move further up the paper than _____ ones.
*Notes*

Slide #29 – The organic solvent moves the proteins up the paper via capillary action. Smaller molecules tend to move further up the paper than larger ones. The retention factor ($R_f$) can be calculated as follows:
*Notes*

$$R_f = \underline{\hspace{8cm}}$$

Slide #30 – The use of _____ to separate proteins was first reported in 1937 by Swedish Biochemist Arne Tiselius, who won the Nobel Prize in Chemistry in 1948. _____ is the migration of ions in an electric field. So for this to work, the dispersed particles must have an electric surface charge, on which an external electric field exerts an electrostatic force.
*Notes*

Slide #31 – In _____ Electrophoresis, the sample is spotted in the middle of a piece of filter paper whose ends are placed in buffer solutions that contain oppositely charged electrodes. The application of a current causes ions to migrate toward the electrode of _____ polarity.

*Notes*

Slide #32 – _____ Electrophoresis. This is one of the most powerful and convenient methods in macromolecular separation. Unlike gel filtration where large molecules traveled faster, here the gel retards the mobility of larger molecules, so smaller molecules go further. _____ electrophoresis produces what is known as an _____. The most common gels are made of polyacrylamide or agarose.

*Notes*

Slide #33 – In _____ Electrophoresis (PAGE), gels are made from the polymerization of acrylamide and methylenebisacrylamide. They are generally not used to separate large molecules (>200 kD). For this you would use an agarose-based gel. However, resolution is usually better with Polyacrylamide Gels. This technique can be used with _____, which is known as SDS-PAGE. SDS is a detergent that binds proteins and causes them to assume a _____ shape. In addition, the large charge that SDS imparts on proteins masks their individual charges. The result of this is that protein separation is based on differences in the _____ of the proteins, and not on differences in shape or charge.

*Notes*

Slide #34 – In a _____, the electrophoretogram is blotted on a sheet of nitrocellulose paper. You then block excess adsorption sites on the nitro-cellulose membrane to prevent _____ binding. This can be done with milk or blocking buffer. The blot is then incubated with the primary antibody, which recognizes the protein of interest. The blot is washed several times, and incubated with a secondary antibody which is covalently linked to an _____ (i.e. HRP). The blot is again washed following the incubation period, and the substrate is added to view the protein band of interest.
*Notes*

Slide #35 – A look at a real Western Blot.
*Notes*

Slide #36 – _____ Polyacrylamide Gel Electrophoresis (SDS-PAGE)-the large charge that SDS imparts on the proteins mask their intrinsic charges. Therefore, they tend to have identical charge-to-mass ratios, and separation is based on _____. _____ molecules are closer to the top, while _____ ones are near the bottom. The gel bands can be detected by staining with _____ blue.
*Notes*

Slide #37 – 2D gel electrophoresis. Here, the proteins are first electrophoresed through a solution having a pH gradient in which the pH smoothly increases. This results in the proteins migrating to the position in the pH gradient that corresponds to their individual _____ point (remember, that is the pH where the protein is neutral). This type of separation is known as _____ Focusing, which is the separation of molecules by their electric charge. Proteins are then further separated by _____.
*Notes*

Slide #38 – A look at the results of a real 2D gel electrophoresis experiment.
*Notes*

Slide #39 – _____ was first described in 1923 by Swedish biochemist Dr. Svedberg. Basically we're talking about speeds in excess of 80,000 rpm (revolutions per minute) or 600,000 g. This has proven to be an invaluable tool for the isolation of proteins, nucleic acids, and subcellular particles.
*Notes*

Slide #40 – DNA staining can be accomplished using _____,
_____, or _____.
*Notes*

Slide #41 – Recommended Problem Sets (from the textbook).
*Notes*

# Chapter 3
## Additional Problem Sets (test format)

1.) Cell lysis is the first step in protein purification. Which of the following could be used to accomplish this?
    A.) High-speed blender.
    B.) French Press.
    C.) Homogenizer.
    D.) Sonicator.
    E.) All of the above are true.

2.) Which of the following is/are true regarding proteins and protein assays?
    A.) Proteins are generally present in very small quantities.
    B.) Assays of proteins have to be highly sensitive to its presence.
    C.) Assays of proteins have to be specific to the protein.
    D.) Assays of proteins have to be relatively convenient to use.
    E.) All of the above are true.

3.) Which of the following is NOT true regarding Immunochemical Techniques?
    A.) Immunochemical techniques can readily detect small amounts of specific proteins.
    B.) Immunochemical techniques are not highly sensitive techniques.
    C.) Immunochemical techniques employs antibodies.
    D.) Antibodies to many proteins are commercially available.
    E.) All of the above are true.

4.) The term "ELISA" stands for:
    A.) Enzyme-Immunosorbent Assay.
    B.) Enzyme-Linked Sorbent Assay.
    C.) Enzyme-Linked Adsorbent Assay.
    D.) Enzyme-Linked Immunosorbent Assay.
    E.) All of the above are true.

5.) When proteins fold, they tend to bury _____ residues in their core.
    A.) Hydrophobic Residues.
    B.) Hydrophilic Residues.

6.) When proteins fold, they tend to do so such that the surface can contain
_____.
    A.) Hydrophobic Residues only.
    B.) Hydrophilic Residues only.
    C.) Both Hydrophobic and Hydrophilic Residues.

7.) Which of the following is true regarding "Salting In"?
   A.) It occurs at relatively low salt concentrations.
   B.) Ions from the salt associate with the surface of the protein.
   C.) The concentration of "free" water is increased.
   D.) The net effect is that the protein becomes more soluble.
   E.) All of the above are true.

8.) Which of the following is true regarding "Salting Out"?
   A.) It occurs at relatively high salt concentrations
   B.) Ions from the salt interact with the solvent.
   C.) The concentration of "free" water is decreased.
   D.) The net effect is that the protein becomes insoluble and precipitates out of solution.
   E.) All of the above are true.

9.) Which of the following is/are true regarding Ion-Exchange chromatography?
   A.) An ion-exchange resin has a ligand with a positive or negative charge.
   B.) Cation exchange chromatography utilizes a negatively charged resin.
   C.) Anion exchange chromatography utilizes a positively charged.
   D.) Proteins that have a charge that is opposite of the resin stick to the column.
   E.) All of the above are true.

10.) Which of the following is NOT true regarding DEAE-Cellulose Resins?
   A.) It is an anion exchanger.
   B.) It is a negatively charged resin.
   C.) It is weakly basic.
   D.) The "DEAE" stands for diethylaminoethyl.
   E.) All of the above are true.

11.) Which of the following is NOT true regarding CM-Cellulose Resins?
   A.) It is a negatively charged resin.
   B.) It is an anion exchanger.
   C.) It is weakly acidic.
   D.) The "CM" stands for carboxymethyl.
   E.) All of the above are true.

12.) Which of the following is true regarding Gel Filtration chromatography?
   A.) It is also referred to as Size Exclusion chromatography.
   B.) It is also referred to as Molecular Sieve chromatography.
   C.) The stationary phase contains hydrated gel beads.
   D.) The gel beads have pores of various sizes.
   E.) All of the above are true.

For questions #13 - #24, consider the following information:

| Protein | p$I$ | Molecular Weight |
|---------|------|------------------|
| A | 1 | 34.5 kDa |
| B | 4.9 | 66.4 kDa |
| C | 5.8 | 340 kDa |
| D | 7.1 | 64.5 kDa |

13.) If you had a mixture of all four proteins listed above, which protein will be eluted from a CM-cellulose ion exchange column first as you gradually increase the salt gradient at pH 7?

    A.) Protein A.
    B.) Protein B.
    C.) Protein C.
    D.) Protein D.
    E.) None of the above.

14.) If you had a mixture of all four proteins listed above, which protein will be eluted from a CM-cellulose ion exchange column last as you gradually increase the salt gradient at pH 7?

    A.) Protein A.
    B.) Protein B.
    C.) Protein C.
    D.) Protein D.
    E.) None of the above.

15.) If you had a mixture of all four proteins listed above, what would be the order in which these proteins are eluted from a CM-cellulose ion exchange column as you gradually increase the salt gradient at pH 7?

    A.) A, B, C, D.
    B.) D, C, B, A.
    C.) C, D, B, A.
    D.) B, A, C, D.
    E.) None of the above.

16.) CM-cellulose resin has a _____ ionizable group.

    A.) Diethylaminoethyl.
    B.) Carboxymethyl.
    C.) Phosphate.
    D.) Methyl sulfonate.
    E.) None of the above.

17.) The use of CM-cellulose resin in a column for protein separation as described in the above questions (#13 - #16) is what type of chromatography?
   A.) Thin Layer Paper chromatography.
   B.) Affinity chromatography.
   C.) Ion Exchange chromatography.
   D.) Gel Filtration chromatography.
   E.) None of the above.

18.) If you had a mixture of protein A and B, which protein will be eluted from a Sephadex G-200 column first?
   A.) Protein A.
   B.) Protein B.
   C.) Protein C.
   D.) Protein D.
   E.) None of the above.

19.) If you had a mixture of protein A and B, which protein will be eluted from a Sephadex G-200 column last?
   A.) Protein A.
   B.) Protein B.
   C.) Protein C.
   D.) Protein D.
   E.) None of the above.

20.) If you had a mixture of protein C and D, which protein will be eluted from a Sephadex G-200 column first?
   A.) Protein A.
   B.) Protein B.
   C.) Protein C.
   D.) Protein D.
   E.) None of the above.

21.) If you had a mixture of protein C and D, which protein will be eluted from a Sephadex G-200 column last?
   A.) Protein A.
   B.) Protein B.
   C.) Protein C.
   D.) Protein D.
   E.) None of the above.

22.) If you had a mixture of all four proteins listed above, what would be the order in which these proteins are eluted from a Sephadex G-200 column?

    A.) D, C, B, A.
    B.) C, B, D, A.
    C.) B, A, C, D.
    D.) C, D, B, A.
    E.) None of the above.

23.) If you had a mixture of all four proteins listed above, which one would have the largest relative elution volume as they are eluted from a Sephadex G-200 column?

    A.) Protein A.
    B.) Protein B.
    C.) Protein C.
    D.) Protein D.
    E.) None of the above.

24.) If you had a mixture of all four proteins listed above, which one would have the smallest relative elution volume as they are eluted from a Sephadex G-200 column?

    A.) Protein A.
    B.) Protein B.
    C.) Protein C.
    D.) Protein D.
    E.) None of the above.

25.) Which of the following is true regarding Affinity chromatography?

    A.) Separation is based on a particular protein's affinity for an immobilized ligand.
    B.) The protein's affinity for the stationary phase is based on tight noncovalent inter-actions.
    C.) The stationary phase contains a ligand that specifically binds the protein of interest.
    D.) The ligand is covalently bound to a solid resin matrix.
    E.) All of the above are true.

26.) Which of the following techniques can potentially be used to release a protein of interest from the column in Affinity chromatography?

    A.) A new elution solution can be added to the column which contains a compound with a higher affinity for the protein-binding site than the bound ligand.
    B.) Changes in pH.
    C.) Changes in ionic strength.
    D.) Changes in temperature.
    E.) All of the above are potential techniques to do this.

27.) Which of the following is true regarding Paper chromatography?
    A.) The paper is the stationary phase.
    B.) An organic solvent (or mixture) is used as the mobile phase.
    C.) A few drops of the sample solution are spotted on the on the paper (~2 cm from the bottom).
    D.) Smaller molecules move further up the paper than larger ones.
    E.) All of the above are true.

28.) The "SDS" in SDS-PAGE stands for _____.
    A.) Sulfur dodecyl sulfate.
    B.) Sodium decyl sulfate.
    C.) Sodium dodecyl sulfate.
    D.) Sulfur decyl sulfate.
    E.) None of the above.

29.) The "PAGE" in SDS-PAGE stands for _____.
    A.) Polyacrylamide Gel Electrophoresis.
    B.) Poly Gel Electrophoresis.
    C.) Postacrylamide Gel Electrophoresis.
    D.) Polyamide Gel Electrophoresis.
    E.) None of the above.

30.) Which of the following can be used to stain DNA in a gel?
    A.) Ethidium Bromide.
    B.) Acridine Orange.
    C.) Proflavin.
    D.) A and C only.
    E.) A, B, and C.

# Chapter 3
## *Answers to Additional Problem Sets (test format)*

1.) E
2.) E
3.) B
4.) D
5.) A
6.) C
7.) E
8.) E
9.) E
10.) B
11.) B
12.) E
13.) A
14.) D
15.) A
16.) B
17.) C
18.) B
19.) A
20.) C
21.) D
22.) B
23.) A
24.) C
25.) E
26.) E
27.) E
28.) C
29.) A
30.) E

# Chapter 4

## *Covalent Structures of Proteins and Nucleic Acids*

*Chapter 4 Summary*:

*Proteins* are commonly referred to as the "building blocks" of life. They can function as enzymes, and can either be enzymatic components or can indirectly influence enzymatic reactions such as hormones. They are involved in the transportation and storage of biological substances. They can coordinate mechanical motion such as in muscle fibers and contractile assemblies. They can process sensory information (i.e. rhodopsin in the retina of your eye). They also play an essential role in the immune system (i.e. immunoglobulins or antibodies) and provide tensile strength to bones, tendons, and ligaments (i.e. collagen). *DNA* on the other hand is commonly referred to as the "genetic archive". As far as various levels of structure, we have primary, secondary, tertiary, and quaternary structures. The primary structure is the amino acid sequence for proteins or base sequences for nucleic acids. The secondary structure involves the spatial arrangements of the backbone atoms. Tertiary structures are the 3-dimensional structure of the entire polypeptide or polynucleotide chain. Quaternary structures involve both noncovalent interactions and disulfide bonds (proteins) that hold two or more subunits together. The first complete amino acid sequence was determined in 1953 by Frederick Sanger. He determined the sequence of bovine insulin. So why is this so important? Well, knowing the amino acid sequence is essential to understanding the protein's mechanism of action, it allows you to compare sequences between proteins which can yield some information related to the function of the protein, and it can give you some clues as to evolutionary relationships between proteins. As far as clinical applications, it is very important because many diseases are caused by mutations. So how do we determine the primary structure? Well, generally speaking we can first prepare the protein for sequencing by cleaving disulfide bonds and possibly separating and purifying unique subunits if necessary. Secondly, we can then sequence the polypeptide chain (using various techniques we will discuss in this chapter). This can generally be done by first digesting subunits to yield smaller peptides, which in some cases can then be directly sequenced. Finally, you can then piece the fragments together to get the entire sequence of the protein (positions of disulfide bonds can then be elucidated based on the positions of cysteine residues). The two major techniques that we will discuss in this chapter regarding N-Terminus identification involve *Dansyl Chloride* and *Edman Degradation*. In the first method, the reagent used is 1-Dimethylamino-napthalene-5-sulfonyl chloride (dansyl chloride). Dansyl chloride reacts with the primary amine of the polypeptide chain under alkaline conditions. The result is a dansyl polypeptide. Acid hydrolysis then liberates the Dansyl amino acid (fluorescent) from the N-Terminus, and free amino acids from the rest of the polypeptide. This is a major disadvantage associated with the use of this technique as continued sequencing is not really possible as the rest of the protein is hydrolyzed into its constituent amino acids. Edman degradation on the other hand is a very useful technique for N-Terminus identification. This technique is named after Pehr Edman, and releases the N-terminal amino acid while leaving the rest of the polypeptide intact, thus allowing for subsequent sequencing. This technique involves the use of phenylisothiocyanate (PITC) which reacts with the N-terminal amino group under mildly alkaline conditions. This results in the formation of a phenylthiocarbamyl (PTC) adduct. Following trifluoroacetic acid (TFA) treatment, the N-terminal residue is cleaved as a thiazoline derivative (the rest of the polypeptide remains intact), which is converted to the more stable phenylthiohydantoin (PTH) derivative. The PTH-amino acid can then be identified by comparing its retention time on an HPLC column (discussed in more detail later) with that of known PTH-amino acids.

PTH-amino acid where "R" is the amino acid side chain.

As for C-Terminus identification, there really is no reliable procedure comparable to Edman Degradation. However, the two major techniques discussed here involve *exopeptidases* (enzymes that cleave terminal residues from polypeptides) and a chemical method referred to as *hydrazinolysis*. The two exopeptidases used for C-Termunis identification discussed here are *carboxypeptidases A and B*. It is important to note that carboxypeptidases exhibit selectivity toward side chains, and therefore rarely reveals more than a few C-terminal residues. For example, while carboxypeptidase B demonstrates selectivity toward basic amino acids such as lysine and arginine, carboxypeptidase A prefers almost any other amino acid (other than proline). It is also important to note that in both cases the neighboring amino acid from the C-Terminus cannot be a proline. In hydrazinolysis, the polypeptide is treated with hydrazine in the presence of a mildly acidic ion exchange resin. The result of this procedure is that all peptide bonds are cleaved resulting in aminoacyl hydrazides of all amino acids with the exception being the C-terminal amino acid, which is released as a free amino acid.

*Disulfide Bonds.* Disulfide bonds must be reduced (or cleaved) between Cysteine residues for a couple of reasons. First, you need to allow for the separation of polypeptide chains. Secondly, you need to prevent the native conformation, which is stabilized by disulfide bonds, from obstructing the proteolytic agents. Two commonly used agents to cleave disulfide bonds are 2-mercaptoethanol and Dithiothreitol. Iodoacetic acid can then be used in order to prevent the reformation of the disulfide bonds. Two mols of 2-mercaptoethanol are needed per disulfide bond, while only one mol of Dithiothreitol is needed per disulfide bond.

*Endopeptidases.* These are proteolytic peptidases that break peptide bonds of nonterminal amino acids. While there are many different types of endopeptidases, the three that you should be familiar with are Trypsin (cleaves after Lys and Arg residues), Chymotrypsin (cleaves after Phe, Trp, and Tyr residues), and Pepsin (cleaves before Phe, Trp, and Tyr residues). Notice that neighboring amino acids again cannot be proline in all cases. While not a peptidase, chemical treatment with Cyanogen Bromide (CNBr) results in the cleavage following methionine residues.

*Amino Acid Analysis.* One very common technique used in the analysis of amino acids is reverse-phase chromatography using HPLC (High-performance liquid chromatography). While normal-phase utilizes a polar stationary phase, reverse-phase employs a nonpolar stationary phase.

*Mass Spectrometry.* Mass Spectrometry (MS) accurately measures the mass-to-charge ratio for ions in the gas phase. The three major ionization sources discussed here are Electrospray ionization (ESI), Matrix-assisted laser desorption/ionization (MALDI), and Fast atom bombardment (FAB). ESI is a technique that was first pioneered by John Fenn. Using this technique, the sample solution is sprayed from a narrow capillary tube that is maintained at high voltage. The solution then readily evaporates from the resultant highly charged droplets. With respect to MALDI, macromolecules are embedded in a crystalline matrix, which is an organic molecule of low molecular mass. Short pulses of intense laser light at a wavelength that is absorbed by the matrix and not the sample, hits the sample. The matrix then ejects the intact macromolecules from its surface into the gas phase due to the energy it absorbs from the laser. Finally, FAB involves dissolving the sample in a low-volatility solvent (i.e. glycerol). The sample is then bombarded with a low energy beam of atoms (i.e. Argon or Xenon). The macromolecules are then ejected into the gas phase. As far as the MS is concerned, while there are many different types, here we focus on time of flight (TOF). This method involves the use of a detector that can determine the mass of macromolecules based on the time it takes for them to travel through its long flight tube. Tandem mass spectrometry in which two mass spectrometers (MS/MS) are coupled in series can also be a valuable tool.

*Peptide Mapping.* This technique takes advantage of the fact that peptide fragments containing amino acid variations will migrate to different positions on the "peptide map" than the corresponding original peptide fragments. Thus, only the variant peptide fragments have to be sequenced rather than the whole protein.

*Nucleic acid sequencing.* The basic strategy for nucleic acid sequencing is similar to protein sequencing. It involves specific degradation and fractionation of polynucleotides. It also involves the ordering of fragments by using degradation procedures that produce fragments with overlapping cleavage points. Restriction Endonucleases enable cleavage of DNA at specific sites.

*Chemical Synthesis of Polypeptides.* Chemical synthesis of polypeptides has considerable biomedical potential. For one thing, you can investigate the properties of polypeptides by systematically varying side chain groups. It is also possible to obtain polypeptides with unique properties (i.e. using nonstandard side chains). Additionally, these techniques allows you to potentially manufacture pharmacologically active polypeptides that are biologically scarce or nonexistent. The first chemical synthesis of a biologically active polypeptide was that of the hormone oxytocin. A very powerful technique used in the chemical synthesis of peptides involves Solid Phase Peptide Synthesis (SPPS). SPPS was first described in 1962 by Bruce Merrifield, and involves the growing peptide chain being covalently anchored to solid support. As the polypeptide chain grows, each new added amino acid (coupling facilitated by coupling reagents) must be protected, and amino acids protected with various protecting groups are commercially available.

*Chemical Synthesis of Oligonucleotides.* This is analogous to that of polypeptide synthesis in that you have a protected nucleotide which is coupled to the growing end of a nucleotide chain. The protecting group is then removed, and the process is repeated. One such method of this is the

Phosphoramidite Method. Here you have the 3' end of an oligonucleotide chain anchored to solid resin. The 5' end is protected with dimethoxytrityl protecting groups, which can be removed with trichloroacetic acid. The coupling reagent that can be used here is tetrazole.

*Chapter 4*
*Lecture Series*

Slide #1 – Introduction
*Notes*

Slide #2 – _____ are commonly referred to as the "building blocks" of life. They function as enzymes, regulators of enzymatic reactions either directly (i.e. enzymatic components) or indirectly (i.e. hormones), and transportation and storage of biological substances. They also coordinate mechanical motion, process sensory information (i.e. _____), and play an essential role in the immune system. Proteins also provide tensile strength to bones, tendons, and ligaments.
*Notes*

Slide #3 – _____ is commonly referred to as the "genetic archive". RNA is integral in the association with protein synthesis. The functions of proteins and nucleic acids can be best understood in terms of their _____.
*Notes*

Slide #4 – Structures. The _____ structure is the amino acid sequence (Protein) or base sequence (Nucleic Acid). The _____ structure is the spatial arrangement of the backbone atoms influenced by interactions between residues that are near each other in the chain (either polypeptide or nucleic acid). The Tertiary structure of the entire poly-(peptide/nucleotide) chain is influenced by interactions between R groups separated by large distances on the chains. The Quaternary structure involves noncovalent interactions and disulfide bonds that hold two or more polypeptide chains (subunits) together.

*Notes*

Slide #5 –
*Notes*

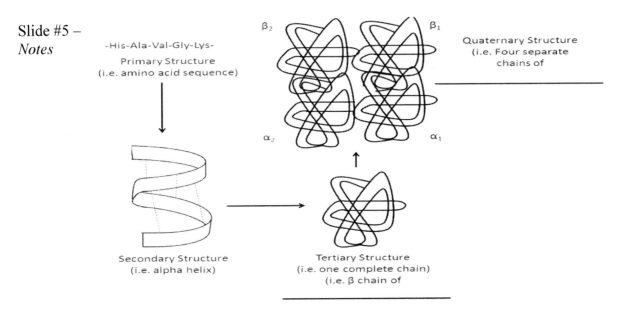

-His-Ala-Val-Gly-Lys-
Primary Structure
(i.e. amino acid sequence)

Secondary Structure
(i.e. alpha helix)

Tertiary Structure
(i.e. one complete chain)
(i.e. β chain of _____

Quaternary Structure
(i.e. Four separate chains of _____

Slide #6 – _____ structures of proteins. The first complete amino acid sequence was determined in 1953 by Frederick Sanger. He determined the sequence of bovine insulin. So why is this so important? Well, the amino acid sequence can yield some insight into understanding _____ of action, allows you to compare sequences which can tell some information related to the function of the protein, and give some clues as to _____ relationships between proteins. As far as clinical applications, it is very important because many diseases are caused by mutations.

*Notes*

Slide #7 – So how do we determine the sequence? First, you prepare the protein for sequencing by cleaving _____ bonds and possibly separating and purifying unique subunits if necessary. You can then sequence the polypeptide chain. This can be done by first digesting subunits to yield smaller peptides. You can then directly sequence the smaller fragments. Again, you can then separate and purify the individual fragments. You then piece the fragments together to get the entire sequence of the protein, and determine the positions of disulfide bonds (if any).
*Notes*

Slide #8 – N-terminal identification. The first is _____ or 1-Dimethylamino-napthalene-5-sulfonyl chloride. It reacts with _____ amines to yield dansylated peptides. The second is _____ degradation which is named after Pehr Edman. It releases the N-terminal amino acid but leaves the rest of the polypeptide intact. This is an important distinction between the two methods.
*Notes*

Slide #9 – Dansyl Chloride. What is the major disadvantage with this technique (look at the products below)?
*Notes*

Slide #10 – _____ Degradation. Is this a better technique for N-Terminal identification compared to Dansyl Chloride? Why? Can you circle the back bone atoms of the N-Terminal residue ($R_1$) in the product here (part of the phenylthiohydantoin ring structure)?
*Notes*

_____ -amino acid

(Phenylthiohydantoin-amino acid)

+ Original polypeptide minus the N-terminal residue

Slide #11 – C-terminus identification. There is no reliable chemical procedure comparable to Edman degradation. However, the C-Terminus amino acid can be identified enzymatically using _____, which are enzymes that cleave terminal residues from polypeptides. _____ specifically are exopeptidases that catalyze the hydrolysis of the C-terminal residues of polypeptides. It should be noted here that because these exhibit selectivity toward side chain groups, they rarely reveal the order of more than a few C-terminal residues.
*Notes*

Slide #12 – While it is not necessary for you to know all of these, the two carboxypeptidases you should know are carboxypeptidase _____ and _____. For carboxypeptidase _____, the C-terminal amino acid cannot be Arg, Lys or Pro (but can be pretty much anything else). For carboxypeptidase _____, the C-terminal amino acid has to be either Lys or Arg. In either case, the neighboring amino acid cannot be Pro.
*Notes*

Slide #13 – Another technique used for C-terminus identification is _____. This involves treating the polypeptide with hydrazine at elevated temperatures for an extended period of time in the presence of a mildly acidic ion exchange resin (this acts as a catalyst). The result is all peptide bonds are cleaved and all of the amino acids exist as aminoacyl _____ except the C-terminal residue.

*Notes*

Slide #14 – Disulfide Bonds. Disulfide bonds must be reduced (or cleaved) between Cysteine residues. This separates the polypeptide chains. Secondly, you need to prevent the native conformation from reforming which would obstruct the proteolytic agents. _____ acid prevents reformation of the disulfide bonds. Can you label the two commonly used agents below that are used to cleave disulfide bonds? How many mols of each are needed per disulfide bond?

*Notes*

$$HS-CH_2-CH_2-OH$$

Name

_____

(need _____ mol(s)/Disulfide Bond)

```
    ┌──SH
    ├──OH
    ├──OH
    └──SH
```

Name

_____

(need _____ mol(s)/Disulfide Bond)

Slide #15 – Prior to amino acid sequence determination, we need to disassociate the subunits present in proteins. This can be done using either acidic or basic conditions, low salt concentrations, elevated temperatures, or denaturing agents such as Urea. We then can purify using a number of techniques (discussed in Chapter 6). The most common being ion exchange and gel filtration (or size exclusion) chromatography.

*Notes*

Slide #16 – Subunit Hydrolysis. Here, you can use acid-catalyzed hydrolysis in which you incubate the polypeptide with 6M HCL. However, you end up destroying Trp residues, and you also end up converting Gln and Asn to Glu and Asp. In base-catalyzed hydrolysis you incubate the polypeptide with 2-4M NaOH. The problem here is that now you have decomposition of Cys, Ser, Thr, and Arg residues. So, this is usually done with a mixture of _____ which results in the complete enzymatic digestion of the polypeptide. _____ are enzymes that catalyzes the hydrolysis of peptides into amino acids.
*Notes*

Slide #17 – The 3 endopeptidases that you should know are _____ (cleaves after Lys and Arg residues), _____ (cleaves after Phe, Trp, and Tyr residues), and _____ (cleaves before Phe, Trp, and Tyr residues).
*Notes*

Slide #18 – A look at Trypsin in action.
*Notes*

Slide #19 – A chemical reagent (not an enzyme) that can be useful is cyanogen bromide (CNBr). It causes specific and quantitative cleavage following _____ residues to from a peptidyl _____ lactone and an aminoacyl peptide.

*Notes*

Slide #20 – An amino acid analyzer can separate amino acids by ion exchange chromatography or reverse-phase HPLC. The amino acids are derivatized so they can be detected. We already talked about Dansyl Chloride (remember you get the Dansyl amino acid which can be detected because it is fluorescent). There is also Edman's reagent where you get the phenylthiohydantoin (PTH) derivative amino acid (or PTH-amino acid), which is compared to the retention times of known amino acids for identification. Also, there is o-pthalaldehyde (OPA) + 2-mercaptoethanol which yields an OPA-derivatized amino acid (fluorescent). What does HPLC stand for?

*Notes*

Slide #21 – A look at OPA-derivatized amino acids coming off a column using RP-HPLC. What is "RP-HPLC" and how is it different from "NP-HPLC"?

*Notes*

Slide #22 – So how can we determine the exact amino acid sequence of polypeptides? Well this can be done using peptide fragments. First, treat the native polypeptide with various endopeptidases and/or chemical reagents (such as CNBr). Purifiy the fragments by HPLC, and then sequence those fragments using Edman degradation. You can than piece together what the native polypeptide sequence was from the various fragments. Finally, you can then assign _____ bond positions.

*Notes*

Slide #23 – Let's look at an example using a 15 amino acid peptide that is treated with both CNBr and Trypsin.

*Notes*

Slide #24 – Now you try one. Let's say that you have a pentapeptide composed of a Gly, 2 Lys, Phe, and Met. When the native peptide undergoes tryptic digestion you get 2 dipeptides and a free Lys. Exposure of the native peptide to CNBr yields a tetrapeptide and a free Gly. One round of Edman Degradation of the native peptide revealed a PTH-Phe residue. What is the order of amino acids?

*Notes*

Slide #25 – Peptide characterization and sequencing from Mass Spectrometry. Mass Spectrometry (MS) accurately measures the mass-to-charge ratio for ions in the gas phase. Prior to 1985 there was really no system available to ionize macromolecules such as polypeptides without destroying them. What made this possible was the advent of three different techniques:

a.) _____ ionization (ESI).
b.) _____ ionization (MALDI).
c.) Fast atom _____ (FAB).
*Notes*

Slide #26 – _____ ionization (or ESI) is a technique that was first pioneered by John Fenn. Using this technique, the sample solution is sprayed from narrow capillary tube that is maintained at high _____ (~4000 V). The solution readily evaporates from the resultant highly charged droplets.
*Notes*

Slide #27 – In _____ (MALDI), macromolecules are embedded in crystalline matrix, which is an organic molecule of low molecular mass. Short pulses of intense _____ light at a wavelength that is absorbed by the matrix and not the sample, hits the sample. The matrix then ejects the intact macromolecules from its surface into the gas phase due to the energy it absorbs from the _____.
*Notes*

Slide #28 – In Fast atom _____ (FAB) mass spec, the sample is dissolved in a low-volatility solvent (i.e. glycerol). The sample is then bombarded with a low energy beam of Argon or Xenon atoms (sometimes Cesium ions). The macromolecules are then ejected into the gas phase.
*Notes*

Slide #29 – Mass spectrometer. They measure the mass with a high degree of accuracy, and there are many types. For example, here you have _____ (TOF) and a Quadrupole.
*Notes*

Slide #30 – MALDI-TOF. Here you have a MALDI head group in which macromolecules are ionized via the short pulses of intense _____ light. The ionized gas macromolecules enter the mass spectrometer (or the "TOF"), which has a detector that can determine the mass of macromolecules based on the time it takes for them to travel through its long flight tube. Hence the term "Time-of-Flight". Obviously, smaller molecules travel _____ than larger ones.
*Notes*

Slide #31 – _____ Mass Spectrometry. This is when you have two mass spectrometers (also called MS/MS) coupled in series. The first MS can select the peptide ion of interest. The selected peptide ion is then passed onto a collision cell where it collides with chemically _____ atoms. The peptide ion then fragments in various locations, and sent on to a second MS, which measures their masses.
*Notes*

Slide #32 – A look at _____ Mass Spectrometry.
*Notes*

Slide #33 – A look at a spectrum and demonstration on how to interpret it.
*Notes*

Slide #34 – So why is it important to determine the peptide sequence? Let's look at a case study here.
*Notes*

Slide #35 – Case study (continued).
*Notes*

Slide #36 – _____ Mapping. Here you digest macromolecules with a sequence specific peptidase. Then, the moderate size fragments are run on an electrophoresis gel. The major advantage of this technique is that it can distinguish between similar but nonidentical proteins (such as mutations). So, if you run the gel, and see that all fragments are the same except one, you can then sequence just that fragment and not the whole protein (i.e. the variant fragment(s)).
*Notes*

Slide #37 – A look at the "fingerprints" of Hemoglobin A (normal).
*Notes*

Slide #38 – A look at the "fingerprints" of Hemoglobin S (sickle cell). This results due to a single amino acid change from _____ to _____.
*Notes*

Slide #39 – Nucleic acid sequencing. The basic strategy for nucleic acid sequencing is similar to protein sequencing. It involves specific degradation and fractionation of polynucleotides. It also involves the ordering of fragments by using degradation procedures that produce fragments with overlapping cleavage points. With proteins we used various endopeptidases. Here, we can use restriction _____ which enable us to cleave DNA at specific sites.
*Notes*

Slide #40 – There are many differences between Nucleic Acid sequencing and Amino Acid sequencing. First, _____ bonds can only be determined with protein sequencing. Modifications made to proteins, which could be essential to their biological function, can only be determined using protein sequencing. This includes modifications such as _____, the addition of _____, and the addition of _____ groups. Nucleic acids specific to the protein of interest are often difficult to isolate. Insertion or deletion of a single nucleotide changes the gene's apparent reading frame and affects predictions for all amino acids thereafter. Also, the standard genetic code is not universal (genetic code for mitochondria and certain protozoa are slightly different).
*Notes*

Slide #41 – Sickle-Cell Trait. People _____ for sickle-cell anemia have sickle-cell trait. Individuals who have the sickle-cell trait lead normal lives (although, erythrocytes have shorter lifetimes). This has linked to _____ resistance. The mosquito-borne protozoan resides in _____, and slightly increases the acidity. This leads to enhanced blood cell sickling, and results in the preferential removal of infected erythrocytes.
*Notes*

Slide #42 – Chemical synthesis of polypeptides has considerable biomedical potential. Can be used to investigate the properties of polypeptides by systematically varying side chain groups. You can also generate polypeptides with unique properties (i.e. nonstandard side chains). You can manufacture pharmacologically active polypeptides that are biologically scarce or nonexistent. The first chemical synthesis of a biologically active polypeptide was that of the hormone _____ (9 amino acids). What is the sequence? Is there a disulfide bond?
*Notes*

Slide #43 – Peptides are generally synthesized from its C-terminus towards its N-terminus (although it can be done the other way). Each amino acid added to the chain has to have its α-amino group blocked/protected to prevent them from reacting with each other (and side chains if applicable). Once added, the N-terminal group must be deprotected so the next amino acid can be added. Every cycle requires a coupling and deblocking step. Peptide synthesis in solution results in very low yields due to the tremendous loss of reagents.
*Notes*

Slide #44 – _____ (SPPS) was first described in 1962 by Bruce Merrifield. In SPPS, the growing peptide chain is covalently anchored (usually by C-terminus) to a _____ support. As the polypeptide chain grows from C- to N-terminus, the α-amino group of each new amino acids must be blocked.
*Notes*

Slide #45 – Another look at the same process.
*Notes*

Slide #46 – Protecting groups.
*Notes*

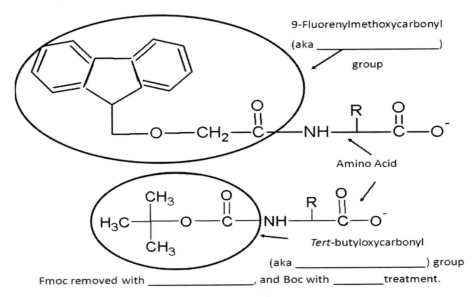

Slide #47 – Coupling reagents. Dicyclohexylcarbodiimide (DCCD) is a commonly used coupling reagent. It forms an acylurea intermediate by binding the C-terminus of the protected amino acid to be added next to the growing polypeptide chain. The acylurea intermediate then readily reacts with the resin bound amino acid.
*Notes*

Slide #48 – Problems and Prospects. A major drawback with peptide synthetic chemistry is the cumulative low yield. Purification can be problematic due to the incomplete and side reactions which liberates closely related products. HPLC alleviates some issues related to purification. Products can be readily assessed through _____.
*Notes*

Slide #49 – Chemical Synthesis of Oligonucleotides. Analogous to that of polypeptide synthesis. A protected nucleotide is coupled to the growing end of a nucleotide chain. The protecting group is then removed, and the process repeated. In the Phosphoramidite Method, the 3' end of oligonucleotide chain is anchored to a solid resin. The 5' end is protected with dimethoxytrityl protecting groups (removed with trichloroacetic acid), and the coupling reagent used is generally tetrazole.
*Notes*

Slide #50 – Recommended Problem Sets (from the textbook).
*Notes*

## Chapter 4
### Additional Problem Sets (test format)

1.) Which of the following is true regarding structures of proteins and nucleic acids?
  A.) The Primary Structure is the amino acid sequence (Protein) or base sequence (Nucleic Acid).
  B.) The Secondary Structure is the spatial arrangement of the backbone atoms influenced by interactions between residues that are near each other in the chain.
  C.) The Tertiary Structure is the 3-D structure of the entire chain.
  D.) The Quaternary Structure involves two or more subunits that can be associated via noncovalent interactions and disulfide bonds.
  E.) All of the above are true.

For questions #2 - #5, consider the following figure:

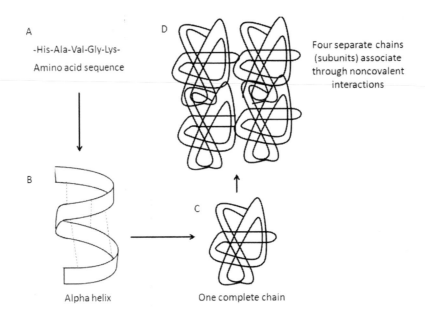

2.) In the above figure, "A" would be considered the _____.
  A.) Primary Structure.
  B.) The Secondary Structure.
  C.) The Tertiary Structure.
  D.) The Quaternary Structure.
  E.) None of the above.

3.) In the above figure, "B" would be considered the _____.
  A.) The Primary Structure.
  B.) Secondary Structure.
  C.) The Tertiary Structure.
  D.) The Quaternary Structure.
  E.) None of the above.

4.) In the above figure, "C" would be considered the _____.
  A.) The Primary Structure.
  B.) Secondary Structure.
  C.) The Tertiary Structure.
  D.) The Quaternary Structure.
  E.) None of the above.

5.) In the above figure, "D" would be considered the _____.
  A.) The Primary Structure.
  B.) Secondary Structure.
  C.) The Tertiary Structure.
  D.) The Quaternary Structure.
  E.) None of the above.

6.) Which of the following can be used to identify the N-Terminal amino acid in a polypeptide chain?
  A.) Dansyl Chloride.
  B.) Carboxypeptidase A.
  C.) Hydrazine (hydrazinolysis).
  D.) Carboxypeptidase B.
  E.) None of the above.

7.) Which of the following can be used to identify the N-Terminal amino acid in a polypeptide chain?
  A.) Carboxypeptidase A.
  B.) Hydrazine (hydrazinolysis).
  C.) Carboxypeptidase B.
  D.) Edman Degradation.
  E.) None of the above.

8.) Which of the following can be used to identify the C-Terminal amino acid in a polypeptide chain?
  A.) Hydrazine (hydrazinolysis).
  B.) Edman Degradation.
  C.) Dansyl Chloride.
  D.) Carboxypeptidase A.
  E.) Both A and D are correct.

9.) Which of the following is commonly-used to cleave disulfide bonds?
  A.) 2-mercaptoethanol.
  B.) Cyanogen Bromide.
  C.) Carboxypeptidase A.
  D.) Dithiothreitol.
  E.) Both A and D are correct.

10.) Which of the following can be (commonly) used following cleavage of disulfide bonds in order to prevent reformation?

    A.) 2-mercaptoethanol.

    B.) Cyanogen Bromide.

    C.) Iodoacetate.

    D.) Dithiothreitol.

    E.) Both A and D are correct.

11.) _____ is an endopeptidase that cleaves a peptide chain following Lys and Arg residues as you go from the N-Terminus to the C-Terminus.

    A.) Pepsin.

    B.) Trypsin.

    C.) Cyanogen Bromide.

    D.) Chymotrypsin.

    E.) None of the above.

12.) _____ is an endopeptidase that cleaves a peptide chain following Phe, Trp, or Tyr residues as you go from the N-Terminus to the C-Terminus.

    A.) Pepsin.

    B.) Trypsin.

    C.) Cyanogen Bromide.

    D.) Chymotrypsin.

    E.) None of the above.

13.) _____ is an endopeptidase that cleaves a peptide chain before Phe, Trp, or Tyr residues as you go from the N-Terminus to the C-Terminus.

    A.) Pepsin.

    B.) Trypsin.

    C.) Cyanogen Bromide.

    D.) Chymotrypsin.

    E.) None of the above.

14.) _____ is a chemical reagent that cleaves a peptide chain following Met residues as you go from the N-Terminus to the C-Terminus.

    A.) Pepsin

    B.) Trypsin.

    C.) Cyanogen Bromide.

    D.) Chymotrypsin.

    E.) None of the above.

15.) Which of the following is true regarding CNBr?
  A.) It causes specific and quantitative cleavage on the C-side of Met residues when used to cleave peptides.
  B.) It is a chemical reagent that cleaves a peptide chain following Met residues as you go from the N-Terminus to the C-Terminus.
  C.) It can be used to activate agarose in affinity chromatography prior to ligand addition.
  D.) When used to cleave peptides, the use of CNBr results in the formation of a peptidyl homoserine lactone and an aminoacyl peptide.
  E.) All of the above are true.

16.) You have a pentapeptide composed of a Gly, 2 Lys, Phe, and Met. When the native peptide undergoes tryptic digestion, you get 2 dipeptides and a free Lys. Exposure of the native peptide to CNBr yields a tetrapeptide and a free Gly. One round of Edman Degradation of the native peptide revealed a PTH-Phe residue. What is the order of amino acids?
  A.) Phe-Lys-Lys-Met-Gly.
  B.) Phe-Lys-Met-Lys-Gly.
  C.) Phe-Met-Lys-Lys-Gly.
  D.) Lys-Phe-Lys-Met-Gly.
  E.) None of the above.

17.) You have a native peptide of unknown amino acid sequence. When the native peptide was treated with trypsin, the sequences of the smaller peptides were as follows:

Val-Ile
Leu-Phe-Arg
Met-Val-Lys

When the native peptide was treated with chymotrypsin, the following sequences were obtained:

Arg-Val-Ile
Met-Val-Lys-Leu-Phe

One round of Edman degradation of the native peptide revealed a PTH-Met residue. What is the order of amino acids of the native peptide?
  A.) Met-Arg-Val-Ile-Val-Lys-Leu-Phe.
  B.) Phe-Met- Val-Lys-Leu-Arg-Val-Ile.
  C.) Met-Phe-Val-Lys-Leu-Arg-Val-Ile.
  D.) Met-Val-Lys-Leu-Phe-Arg-Val-Ile.
  E.) None of the above.

18.) You have a native peptide of unknown amino acid sequence. When the native peptide was treated with trypsin, the sequences of the smaller peptides were as follows:

Gly-Tyr-Met-Phe
Val-Leu-Met-Arg
Asn-Thr-Met-Lys

When the native peptide was treated with Cyanogen Bromide, the following sequences were obtained:

Phe
Val-Leu-Met
Lys-Gly-Tyr-Met
Arg-Asn-Thr-Met

One round of Edman degradation of the native peptide revealed a PTH-Val residue. What is the order of amino acids of the native peptide?
A.) Val-Leu-Met-Met-Arg-Asn-Thr-Lys-Gly-Tyr-Met-Phe.
B.) Val-Leu-Met-Lys-Gly-Tyr-Arg-Asn-Thr-Met-Met-Phe.
C.) Val-Leu-Met-Arg-Asn-Thr-Met-Lys-Gly-Tyr-Met-Phe.
D.) Val-Met-Leu-Arg-Asn-Thr-Met-Lys-Gly-Tyr-Met-Phe.
E.) None of the above.

19.) When using mass spectrometry, which of the following can be used to produce gas phase ions in the ionization process?
A.) ESI.
B.) MALDI.
C.) FAB.
D.) Only A and B are correct.
E.) All of the above are true (A, B, and C).

20.) When using mass spectrometry, what does MALDI stand for?
A.) Matrix-Associated Laser Desorption Ionization.
B.) Matrix-Allowed Laser Desorption Ionization.
C.) Matrix-Assisted Laser Desorption Ionization.
D.) Matrigel-Assisted Laser Deportation Ionization.
E.) None of the above.

21.) When using mass spectrometry, what does ESI stand for?
A.) Electrospit Ionization.
B.) Electrospray Ionization.
C.) Extra Spray Ionization.
D.) Electro Spill Ionization.
E.) None of the above.

22.) When using mass spectrometry, what does FAB stand for?
    A.) Flowing Atom Bombardment.
    B.) Fast Acting Bombardment
    C.) Fast Atom Bombardment.
    D.) Flowing Acting Bombardment.
    E.) None of the above.

23.) When using mass spectrometry, which of the following techniques involves dissolving the sample in a low-volatility solvent such as glycerol, and bombarding the sample with a low energy beam of atoms (usually Ar or Xe)?
    A.) ESI.
    B.) MALDI.
    C.) FAB.
    D.) FAT.
    E.) None of the above.

24.) When using mass spectrometry, which of the following techniques involves embedding the sample in a crystalline matrix and irradiating it with short pulses of intense laser light.
    A.) ESI.
    B.) MALDI.
    C.) FAB.
    D.) FAT.
    E.) None of the above.

25.) When using mass spectrometry, which of the following techniques involves the sample being sprayed from a narrow capillary tube which is maintained at high voltage?
    A.) ESI.
    B.) MALDI.
    C.) FAB.
    D.) FAT.
    E.) None of the above.

26.) When using mass spectrometry, what does MALDI-TOF stand for?
    A.) Matrix-Assisted Laser Desorption Ionization Time-of-Fire.
    B.) Matrix-Assisted Laser Desorption Ionization Time-of-Flight.
    C.) Matrix-Associated Laser Desorption Ionization Time-of-Fire.
    D.) Matrix-Associated Laser Desorption Ionization Time-of-Flight.
    E.) None of the above.

27.) When using mass spectrometry, which of the following is true regarding the use of a tandem mass spectrometer (MS/MS)?
- A.) A tandem mass spectrometry has two mass spectrometers (MS/MS) coupled in series.
- B.) The first MS can be used to select the peptide ion of interest.
- C.) The selected peptide ion can then be passed onto a collision cell.
- D.) The second MS can then be used to measure the masses of the resultant   fragments from the collision cell.
- E.) All of the above are true.

28.) Which of the following is true regarding peptide mapping?
- A.) This technique is also known as "fingerprinting".
- B.) This technique takes advantage of the fact that peptide fragments containing amino acid variations will migrate to different positions on the "peptide map" than the corresponding original peptide fragments.
- C.) Only the variant fragments need to be elucidated and sequenced.
- D.) The major advantage here being that the whole protein does not necessarily need to be sequenced (only variant peptides).
- E.) All of the above are true.

29.) In sickle-cell anemia, the negatively charged glutamic acid residue is replaced by the neutral amino acid _____.
- A.) Glycine.
- B.) Isoleucine.
- C.) Leucine.
- D.) Valine.
- E.) None of the above.

30.) Hemoglobin S, the variant responsible for the misshapen red blood cells characteristic of the disease sickle-cell anemia, is potentially advantageous to heterozygotes because it confers some level of resistance to _____.
- A.) Rickets.
- B.) AIDS.
- C.) Polycythemia minor.
- D.) Malaria.
- E.) None of the above.

## Chapter 4
### *Answers to Additional Problem Sets (test format)*

1.) E
2.) A
3.) B
4.) C
5.) D
6.) A
7.) D
8.) E
9.) E
10.) C
11.) B
12.) D
13.) A
14.) C
15.) E
16.) A
17.) D
18.) C
19.) E
20.) C
21.) B
22.) C
23.) C
24.) B
25.) A
26.) B
27.) E
28.) E
29.) D
30.) D

# Chapter 5

# *Three-Dimensional Structures of Proteins*

*Chapter 5 Summary*:

*The Peptide Group.* X-ray studies done in the early to mid 1900s have revealed that the peptide group has a rigid, planar structure. The rigidity of the structure can be attributed to the double-bond character of the peptide bond (~40% double-bond character). The peptide's C-N bond (from the amide bond) is shorter (~1.33 A) than a regular C-N single bond (~1.47 A). Peptide groups usually assume the trans conformation in which each alpha-Carbon is on opposite sides of the amide bond (see figure below).

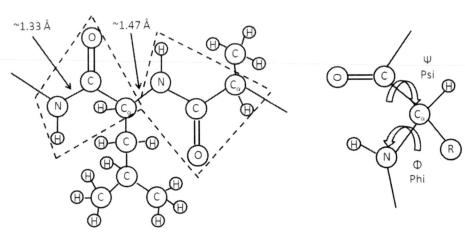

Figure depicting the backbone of a polypeptide chain (enclosed within dotted lines). Amino acids shown are Leucine and Alanine (left to right). Notice that the R groups are not part of the backbone.

Figure depicting Phi and Psi torsion angles.

The cis conformation is less stable than the trans conformation because of the steric interference of the R groups or side chains. Pro can be an exception (~10% of Pro residues in a protein follow a cis peptide bond). The peptide backbone is defined as the atoms that participate in peptide bonds (ignoring side chains).

*Torsion angles.* The conformation of the backbone can be described by torsion angles (aka dihedral angles or rotational angles). The name of the torsion angle around the alpha carbon - nitrogen bond is derived from the Greek alphabet and is referred to as "Phi", while the torsion angle around the alpha carbon - carbonyl bond known as "Psi" (see figure above). There are several steric constraints on these torsion angles which can limit the structure they adopt depending on the identities of the "R" groups. It is helpful when studying torsion angles to remember the various confirmations of ethane from organic chemistry (staggered and eclipse), with staggered being most stable. The concept here is similar.

*Ramachandran Diagram.* It is important to note that the angles that Phi and Psi can adopt are very limited due to sterically forbidden conformations. Sterically allowed values for various secondary structures of proteins have been calculated and can be depicted in a diagram called a Ramachandran diagram (see figure below) named after the inventor G.N. Ramachandran. Notice that most of the diagram is blank. That is because almost 75% of the Ramachandran diagram, which is to say most combinations of Phi and Psi, are conformationally inaccessible to most polypeptides. In any event, clustering within certain ares of the diagram allows us to predict the secondary structure. While there are many possibilities here, you are only responsible for parallel and antiparallel β pleated sheets, right-handed α helices, and left-handed α helices (see figure below).

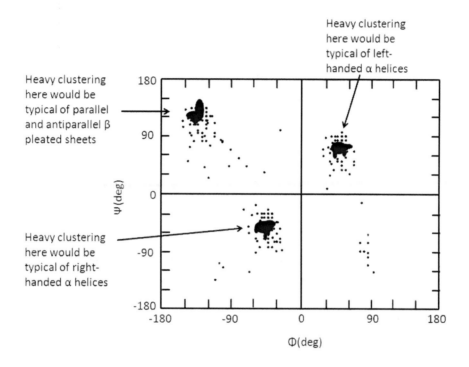

*Helical Structures.* The polypeptide chain is twisted by the same amount about its alpha Carbon atoms. Instead of characterizing helices with phi and psi torsion angles, they can be characterized by the number, n, of peptide units per helical turn, and by its pitch, p, which is the distance the helix rises along its axis per turn. Traditionally, right-handed helices are designated a positive number for n, while left-handed helices have a negative number to represent n.

*The α helix* was discovered by Linus Pauling in 1951, and is seen as one of the landmarks in structural Biochemistry. The α helix is the only helical polypeptide arrangement that has both allowed conformation angles and a favorable hydrogen bonding pattern. The "glue" that holds polypeptides together is hydrogen bonds. Obviously this is not unique to helical structures, as hydrogen bonds play an essential role in various other secondary structures. The arrangement of the polypeptide chain results in a particularly rigid structure. The α helix is a common secondary structure found in both fibrous and globular proteins. Polypeptides made from L-α-amino acid residues form right-handed α helices, while polypeptides made from D-α-amino acid residues form left-handed α helices.

*Beta (β) Structures.* β pleated sheets are similar to the alpha helix in that they also have repeating Phi and Psi angles that fall within the allowable region of Ramachandran diagrams. They also use the full hydrogen bonding capacity of the peptide backbone. However, unlike α helices, hydrogen bonding occurs between neighboring chains (the α helix has hydrogen bonding within one chain). There are two kinds of β pleated sheets that we discuss here, specifically parallel and antiparallel pleated β sheets. In parallel pleated β sheets, neighboring hydrogen bonded polypeptide chains extend in the same direction, while in antiparallel pleated β sheets neighboring hydrogen bonded polypeptide chains run in opposite directions.

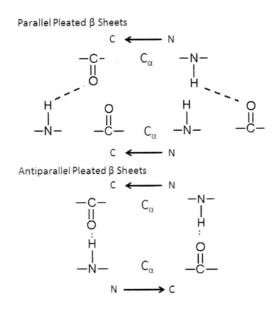

β sheets are very common structural motifs in proteins. They can consist of 2 to 22 polypeptide strands (6 being the average). They are known to be up to 15 residues long (and again, 6 being the average). Another important point here is that parallel β sheets tend to be less stable than antiparallel as parallel β sheets are distorted when compared to antiparallel (look at the hydrogen bonding pattern above).

*Nonrepetitive Structures.* Approximately half of the polypeptide segments found in a protein have a coil or a loop confirmation (the other half being helices and β sheets). They are irregular and therefore difficult to describe, however, nonrepetitive structures have an order associated with them. The two nonrepetitive structures discussed here are β bends (aka reverse turns) and omega loops. *β bends* normally connect successive strands of antiparallel β sheets (hence the name). They almost always occur at protein surfaces Most β bends consist of four amino acid residues that are classified as one of two types that differ by a 180° flip of residues 2 and 3 (see figure below). They are designated type I and type II β bends. While residue two is often Pro for both types, residue three in type II β bends is generally Gly as the oxygen from residue two crowds the residue three R group.

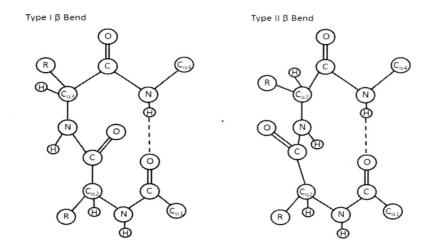

Type I β Bend          Type II β Bend

*Omega loops* contain 6 to 16 amino acid residues. The end-to-end distance is less than 10 Angstroms. They are also present at the protein surface and they get their name because they look like the "omega" symbol.

*Fibrous proteins* are highly elongated molecules. They generally play a protective, connective, or supportive role (i.e. in the skin, tendons, and bone). Some fibrous proteins have a motive function (i.e. muscle). Two common and well-characterized proteins that are discussed here include keratin and collagen. *Keratin* is a mechanically durable protein, and chemically unreactive. They are characterized as being either α keratins (α is used because they are most representative of α helices), which occur in mammals, or β keratins (β is used because they are most representative of β sheets), which occur in birds and reptiles. Mammals carry over 30 keratin genes whose products are either relatively acidic (Type I) or relatively basic (Type II) polypeptides. α keratin polypeptides form closely associated pairs of α helices (one acidic and one basic). The two keratin chains are twisted in a parallel left-handed coil, and therefore the assembly of keratin is said to have a coiled coil. Each polypeptide chain has a heptad (7-residue) pseudorepeat, with hydrophobic amino acids at various locations that lines up with the other associated α helix such that a hydrophobic strip is formed. Keratin is rich in Cys residues, which allows for the formation of many disulfide bonds between adjacent polypeptide chains. Thus, α keratin is resistant to stretching. Keratins can either be "hard" (for example hair and nails) or "soft" (like skin) depending on sulfur content. Our hair is a good example of the macro-scopic organization of α keratins. This explains why in perms it is first necessary to reductively cleave these disulfide bonds with mercaptans (remember we talked about 2-mercaptoethanol, and how it is used to break disulfide bonds in chapter 7), curl the hair, then reestablish the disulfide bonds to hold in place. *Epidermolysis Bullosa (EBS)*. EBS develops due to the rupturing of epidermal cells resulting in skin blistering. It can be caused by mutation of the keratin gene. There is a book entitled "The boy whose skin fell off", in which a boy is portrayed who has this condition. The other fibrous protein discussed here is collagen. *Collagen* is a triple helical structure, and is the most abundant protein in vertebrates. It is organized into water-insoluble fibers and is of great strength. Thus, it is not surprising that it is found in bone and connective tissue. There are at least 20 distinct collagen types that occur in different tissues of the same individual. Collagen has a distinct amino acid composition (~1/3 of the residues are Gly, and another ~30% are Pro and Hyp. A common motif found in collagen is the Gly-X-Y motif, where X and Y can be any amino acid. In many instances X and Y are Pro and Hyp respectively. Gly allows for tight turns, while Pro and Hyp allow for rigidity. Hyp also confers stability to collagen due to hydrogen bonding. The hydroxylated residues appear after the polypeptide is synthesized by an enzyme known as prolylhydroxylase, which requires ascorbic acid (vitamin C) for its activity. Scurvy can result from a deficiency of vitamin C. All known amino acid changes within Type I collagen's triple helix region result in abnormalities. For example, Osteogenesis Imperfecta is a rare heritable disorder that results from the mutation of Type I collagen (the major structural protein in most human tissues). This is also commonly referred to as "brittle bone disease".

*Globular proteins* exist as compact spheroidal molecules. Examples of globular proteins include enzymes, as well as transport and receptor proteins. Globular proteins may contain both α helices and β sheets. In fact, on average, most globular proteins contain ~30 % α helices and slightly less β sheets. A quick look at the enzyme Carbonic Anhydrase reveals the various α helices and β sheets present within the molecule. Carbonic Anhydrase is an enzyme that converts carbon dioxide to bicarbonate to maintain acid-base balance in blood and other tissues, and to help transport carbon dioxide out of tissues. It is a metalloenzyme, which means it requires a metal in the active site for its activity (in this case a Zinc ion). The zinc ion is held into place by the imidazole ring of 3 Histidine residues.

*Supersecondary Structures.* While there are various supersecondary structures, three are discussed here. The βαβ motif (see figure below) is the most common. The β hairpin motif is another common motif. Finally, notice the antiparallel α helices in the αα motif that arrange themselves to allow for side chains to interdigitate.

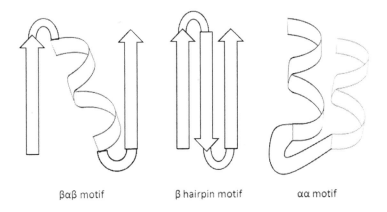

βαβ motif          β hairpin motif          αα motif

*Nuclear Magnetic Resonance spectroscopy (NMR).* You have many types of NMR. One that you might be familiar with from Organic Chemistry is COSY, or correlation spectroscopy. This will tell you how close hydrogen atoms are through bonds. In this case, this is an example of NOESY, which stands for Nuclear Overhauser Effect Spectroscopy. The Nuclear Overhauser Effect (NOE) determines interproton distances through space. You should be able to interpret NOESY data (see your notes, Slide #42).

## *Chapter 5*
### *Lecture Series*

Slide #1 – Introduction
*Notes*

Slide #2 – The _____ group has a rigid, planar structure, which can be attributed to the _____-bond character of the peptide bond (~40% double-bond character). The peptide's Carbonyl-Nitrogen bond is shorter than a regular Carbon -Nitrogen single bond.
*Notes*

Slide #3 – Peptide groups usually assume the _____ conformation in which each alpha-carbon is on opposite sides of the amide bond. The cis conformation is less stable than the trans conformation because of the steric interference of the R groups or side chains. _____ can be an exception. The backbone is defined as the atoms that participate in peptide bonds (ignoring side chains).
*Notes*

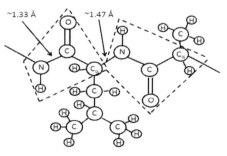

Slide #4 – Torsion angles. The conformation of the backbone can be described by torsion angles (dihedral angles or rotational angles). The name of the torsion angle around the $C_\alpha$-N bond is derived from the Greek alphabet and is known as "_____", while the torsion angle around the $C_\alpha$-C (carbonyl) bond derived from the Greek alphabet is known as "_____". There are several steric constraints on these torsion angles which can limit the structure they adopt depending on the identities of the "R" groups.
*Notes*

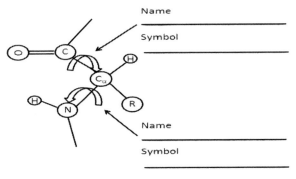

Slide #5 – Conformations of _____. There is limited rotation about the C-C bond. What are the two conformations here? Which is more stable?
*Notes*

Slide #6 – _____ Diagram. The angles that Phi and Psi can adopt are also very limited due to sterically forbidden conformations. Sterically allowed values for various secondary structures of proteins have been calculated and depicted in a diagram called a _____ diagram. Almost 75% of the diagram is blank, which is to say that most combinations of Phi and Psi is conformationally inaccessible to most polypeptides.
*Notes*

Slide #7 – Example of a Sterically Forbidden Conformation. Notice the collision between the carbonyl oxygen and the following amide hydrogen.
*Notes*

Slide #8 – Can you label the following diagram?
*Notes*

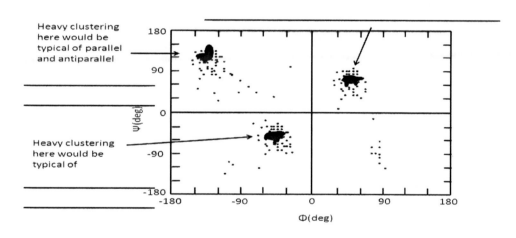

Slide #9 – This Ramachandran diagram represents a protein that probably consists mostly of right handed helices, left handed helices, poly-L-proline II helices (which is probably not the right answer as this is not one of the three that you are responsible for), or Beta sheets?
*Notes*

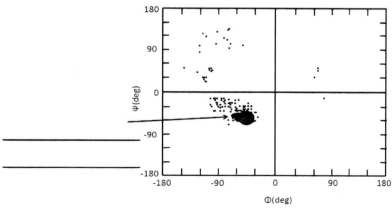

Slide #10 – Ramachandran Diagram of Glycine. Glycine residues have far greater conformational freedom than other amino acids.

*Notes*

Slide #11 – Helical Structures, the polypeptide chain is twisted by the same amount about its alpha Carbon atoms. Instead of characterizing helices with Phi and Psi torsion angles, they can be characterized by the number, n, of peptide units per helical turn, and by its pitch, p, which is the _____ the helix rises along its axis per turn.

*Notes*

Slide #12 – Helical Structures (continued). Notice that in right-handed helices, "n" is a positive number, and in left-handed helices "n" is a negative number.

*Notes*

Slide #13 – The α helix was discovered by Linus Pauling in 1951, and is seen as one of the landmarks in structural Biochemistry. The α helix is the only helical polypeptide arrangement that has both allowed _____ angles and a favorable _____ bonding pattern. The "glue" that holds polypeptides together is, in part, _____bonds. Obviously, this is not unique to helical structures as these bonds play an essential role in various other secondary structures.
*Notes*

Slide #14 – The α helix (continued). The arrangement of the polypeptide chain results in a particularly rigid structure. The α helix is a common secondary structure found in both fibrous and globular proteins. Polypeptides made from _____-α-amino acid residues form right-handed α helices, while polypeptides made from _____-α-amino acid residues form left-handed α helices.
*Notes*

Slide #15 – Beta (β) pleated sheets. Similar to the alpha helix in that they also have repeating Phi and Psi angles that fall within the allowable region of Ramachandran diagrams. They use the full hydrogen bonding capacity of the peptide backbone. However, unlike α helices, hydrogen bonding occurs between _____ chains (the α helix has hydrogen bonding within one chain). There are two kinds of β pleated sheets that we are going to talk about. They are _____ and Antiparallel pleated β sheets.
*Notes*

Slide #16 – _____ pleated β sheets.
*Notes*

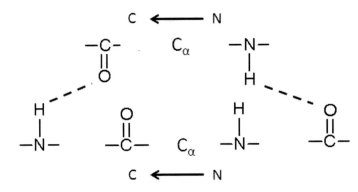

Slide #17 – _____ pleated β sheets.
*Notes*

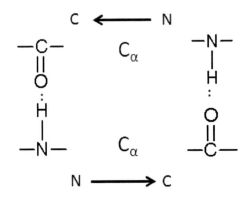

Slide #18 – Can you tell which one this is?
*Notes*

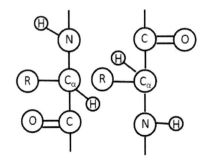

Beta-Sheet

_____

Slide #19 – Beta (β) sheets (continued). They are very common structural motifs in proteins. They can consist of 2 to 22 polypeptide strands (6 being the average), and are known to be up to 15 residues long (6 again being the average). _____β sheets tend to be less stable than _____. This is due to the fact that parallel β sheets are _____ when compared to antiparallel.

*Notes*

Slide #20 – Nonrepetitive structures. Approximately half of the polypeptide segments found in a protein have a coil (NOT RANDOM COIL) or a loop confirmation (the other half being helices and β sheets). They are irregular and therefore difficult to describe, however, nonrepetitive structures are ordered. Here we discuss two structures. Specifically, we discuss _____ (Reverse turns), and _____ loops.

*Notes*

Slide #21 – Beta (β) bends normally connect successive strands of _____β sheets. They almost always occur at protein surfaces. Most β bends consist of four amino acid residues that are classified as either one of two major types that differ by a 180° flip of residues 2 and 3. They are designated type I and type II β bends.

*Notes*

Slide #22 – A look at type I and II β bends. In Type II β bends, the oxygen atom of residue 2 crowds the $C_\beta$ atom of residue 3. Residue 2 is usually _____ for both types.

*Notes*

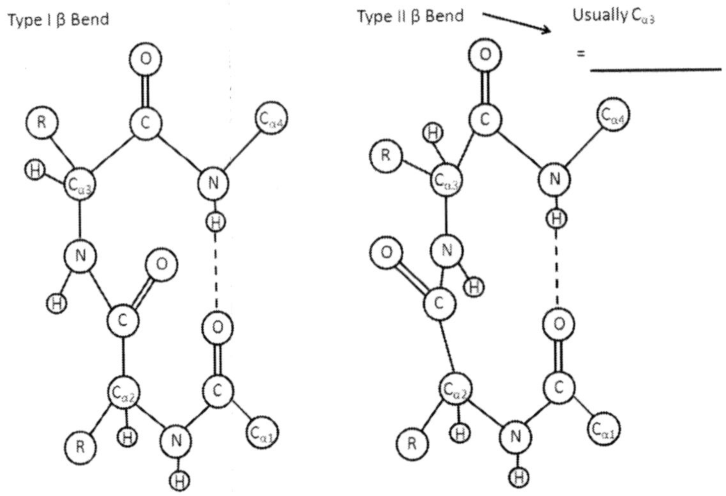

Slide #23 – _____ loops. They contain 6 to 16 amino acid residues. The end-to-end distance is less than 10 Angstroms. They are also present at the protein surface and they get their name because they look like the "_____" symbol.

*Notes*

Slide #24 – Fibrous proteins. Fibrous proteins are highly elongated molecules. They generally play a protective, connective, or supportive role (i.e. in the skin, tendons, and bone). Some fibrous proteins have a motive function (i.e. muscle). The two common well-characterized proteins discussed here are _____ and _____.

*Notes*

Slide #25 – _____ is a mechanically durable protein, and chemically unreactive. The two kinds of keratins are the α keratins (α is used as they are most representative of α helices) and they occur in mammals. The β keratins (β is used as they are most representative of β sheets) occur in birds and reptiles. Mammals carry over 30 keratin genes whose products are either relatively acidic (Type I) or relatively basic (Type II) polypeptides. The α keratin polypeptides form closely associated pairs of α helices.
*Notes*

Slide #26 – Keratin (continued). Each pair consist of a Type I and a Type II keratin chain. The two keratin chains are twisted in a parallel left-handed coil. The assembly of keratin is said to have a _____ _____ because each α helix in itself forms a helical structure. Each polypeptide chain has a _____ (7-residue) pseudorepeat (such that certain amino acids line-up to form a hydrophobic strip).
*Notes*

Slide #27 – Keratin (continued). Certain amino acids line-up to form a hydrophobic strip).
*Notes*

Slide #28 – Keratin (continued). Keratin is rich in Cys residues, which form many disulfide bonds that cross link adjacent polypeptide chains (thus, it is resistant to stretching). Keratins can either be "Hard" (for example hair and nails) or "Soft" (like skin) depending on sulfur content. In perms, hair is reductively cleaved with mercaptans, curled, and then disulfide bonds can be reestablished by the application of an oxidizing agent.

*Notes*

Slide #29 – Keratin (continued). _____ formed by two staggered and antiparallel rows of associated head-to-tail coiled coils. _____ are formed from dimerized Protofilaments. _____ formed from 4 Protofibrils.

*Notes*

Slide #30 – Keratin (continued). _____ contain several Microfibrils in the macroscopic organization of hair.

*Notes*

Slide #31 – _____ (EBS). This disease develops due to the rupturing of epidermal cells resulting in skin blistering. It can be caused by mutation of the keratin gene, and amino acid sequence abnormalities may be present in either keratin 14 or keratin 5 (which are the dominant Type I and Type II keratins in these cells).
*Notes*

Slide #32 – The other fibrous protein discussed here is _____, which occurs in all multicellular animals. It is an extracellular protein, and is the most abundant protein in vertebrates. They are organized into insoluble fibers that have incredible tensile strength. It is therefore the major stress-bearing component of connective tissue.
*Notes*

Slide #33 – There are at least 20 distinct collagen types that occur in different tissues of the same individual. It has a very distinct amino acid composition with high levels of Gly, Pro, and Hyp (Hydroxyproline). Can you draw a 4-Hydroxyprolyl residue?
*Notes*

Structure of a 4-Hydroxyprolyl
residue (Hyp)

Slide #34 – _____ residues confers stability to collagen because it offers more hydrogen bonding to hold the triple helix together. The hydroxylated residues appear after the polypeptide is synthesized by _____, which is an enzyme that requires ascorbic acid (vitamin C) for its activity.
*Notes*

Slide #35 – _____ is a disease that results from an ascorbic acid deficiency resulting in skin lesions, blood vessel fragility etc. Collagen synthesized without _____ residues are denatured at~ 25°C (compared to 39°C).
*Notes*

Slide #36 – Collagen ("A Triple Helical Cable") consists of repeating triplets of Gly-X-Y amino acids, where X is commonly _____ and Y is a _____ residue. Prolylhydroxylase is specific to the _____ position.
*Notes*

Slide #37 – Collagen's three polypeptide chains are parallel and wind around each other with a _____-handed rope-like twist to form a triple helical structure. It is well-packed, rigid, and the triple helical structure is responsible for its characteristic tensile strength.
*Notes*

Slide #38 – _____ is also called brittle bone disease. It is a rare heritable disorder that results from the mutation of Type I collagen (the major structural protein in most human tissues). All known amino acid changes within Type I collagen's triple helix region result in abnormalities.
*Notes*

Slide #39 – Globular proteins. They exist as compact spheroidal molecules. Examples of globular proteins include enzymes, as well as transport and receptor proteins. Globular proteins may contain both α helices and β sheets. In fact, on average, most globular proteins contain ~30 % α helices and 30% β sheets.
*Notes*

Slide #40 – _____ Anhydrase is an enzyme that converts carbon dioxide to bicarbonate to maintain acid-base balance in the blood and other tissues, and to help transport carbon dioxide out of tissues. It is a metalloenzyme, which means it requires a metal in the active site for its activity (in this case a Zinc ion). The zinc ion is held into place by the _____ ring of 3 _____ residues.

*Notes*

Slide #41 – Supersecondary Structures. Can you label them below?
*Notes*

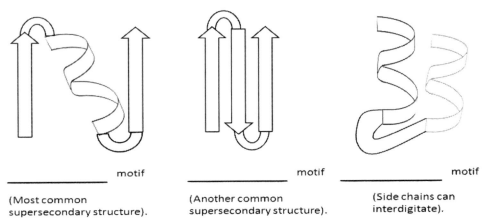

_____ motif          _____ motif          _____ motif

(Most common              (Another common              (Side chains can
supersecondary structure).  supersecondary structure).   interdigitate).

Slide #42 – NOESY, which stands for Nuclear Overhauser Effect Spectroscopy. The Nuclear Overhauser Effect (NOE) determines interproton distances through _____.
*Notes*

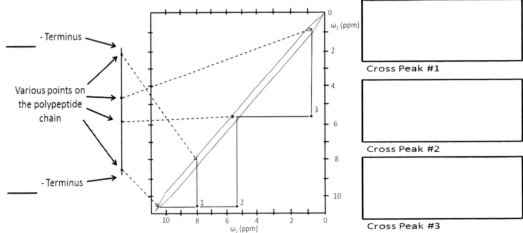

Slide #43 – Structural Bioinformatics Websites.
*Notes*

Slide #44 – Recommended Problem Sets (from the textbook).
*Notes*

## Chapter 5
### Additional Problem Sets (test format)

1.) The rigidity of the peptide group in a protein can in part be attributed to the _____ character of the peptide bond.
    A.) Single-bond.
    B.) Double-bond.
    C.) Triple-bond.
    D.) Ionic bond.
    E.) None of the above.

2.) Peptide groups usually assume the _____ conformation (i.e. each $C_\alpha$ is on the _____ sides of amide bond that join them).
    A.) Cis; Same.
    B.) Trans; Same.
    C.) Cis; Opposite.
    D.) Trans; Opposite.
    E.) None of the above.

3.) The _____ of polypeptides refer to the atoms that participate in peptide bonds, while ignoring the individual side chains.
    A.) "R" groups.
    B.) Back.
    C.) Backbone.
    D.) Bone.
    E.) None of the above.

4.) The conformation of the polypeptide backbone can be described by examining the rotation about the $C_\alpha$-N bond and the $C_\alpha$-C bond. These rotational angles are referred to as _____ angles.
    A.) Torsion.
    B.) Dihedral.
    C.) Rotation.
    D.) A, B, and C are correct.
    E.) None of the above.

5.) The torsion angle around the $C_\alpha$-N bond is the _____ angle, while the torsion angle around the $C_\alpha$-C bond is the _____ angle.
    A.) Phi ($\Phi$); Psi ($\Psi$).
    B.) Psi ($\Psi$); Phi ($\Phi$).
    C.) Phi ($\Psi$); Psi ($\Phi$).
    D.) Psi ($\Phi$); Phi ($\Psi$).
    E.) None of the above.

6.) In class, we looked at a conformation map of sterically allowed conformations of Φ and Ψ angles depicted in a _____ diagram.
    A.) Rachael.
    B.) Roman.
    C.) Ramachandran.
    D.) Reuben.
    E.) None of the above.

7.) Consider the Ramachandran Diagram that you have in your notes (Slide#8) and the secondary structures that you are responsible for in this class. Heavy clustering in the upper right-hand quadrant of the diagram would probably be indicative of what secondary structure?
    A.) Right-handed α helix.
    B.) Left-handed α helix.
    C.) A parallel Reuben helix.
    D.) β pleated sheets (either parallel or antiparallel).
    E.) None of the above.

8.) Consider the Ramachandran Diagram that you have in your notes (Slide#8) and the secondary structures that you are responsible for in this class. Heavy clustering in the upper left-hand quadrant of the diagram would probably be indicative of what secondary structure?
    A.) Right-handed α helix.
    B.) Left-handed α helix.
    C.) A parallel Reuben helix.
    D.) β pleated sheets (either parallel or antiparallel).
    E.) None of the above.

9.) Consider the Ramachandran Diagram that you have in your notes (Slide#8) and the secondary structures that you are responsible for in this class. Heavy clustering in the lower left-hand quadrant of the diagram would probably be indicative of what secondary structure?
    A.) Right-handed α helix.
    B.) Left-handed α helix.
    C.) A parallel Reuben helix.
    D.) β pleated sheets (either parallel or antiparallel).
    E.) None of the above.

10.) All of the information necessary for folding the peptide chain into its "native" structure is contained in the _____ of the peptide.
    A.) Amino acid composition.
    B.) Amino acid sequence.
    C.) Amino acid side chains.
    D.) Configuration.
    E.) None of the above.

11.) Which of the following is/are true regarding helical structures?
   A.) The polypeptide chain is twisted by the same amount about its $C_\alpha$ atoms.
   B.) A helix may be characterized by the number, n, of peptide units per helical turn.
   C.) In right-handed helices n is a positive number, and in left-handed helices n is a negative number.
   D.) A helix may also be characterized by its pitch, p, the distance the helix rises along its axis per turn.
   E.) All of the above are true.

12.) Which of the following is/are true regarding the α helix?
   A.) The α helix is the only helical polypeptide arrangement that has both allowed conformation angles and favorable hydrogen bonding pattern.
   B.) Hydrogen bonds are important intermolecular forces in the α helix.
   C.) The arrangement of the polypeptide chain results in a particularly rigid structure.
   D.) The α helix is a common secondary structure found in both fibrous and globular proteins.
   E.) All of the above are true.

13.) Which of the following is/are true regarding β pleated sheets?
   A.) β pleated sheets have repeating $\Phi$ and $\Psi$ angles that fall within the allowable region of Ramachandran diagrams.
   B.) Hydrogen bonding occurs between neighboring chains.
   C.) β sheets are common structural motifs in proteins.
   D.) Parallel β sheets tend to be less stable than antiparallel sheets.
   E.) All of the above are true.

14.) In Type II β bends, the $C_{\alpha 2}$ residue is generally _____, while the $C_{\alpha 3}$ residue is generally _____.
   A.) Glycine; Proline.
   B.) Tryptophan; Proline.
   C.) Proline; Glycine.
   D.) Proline; Tryptophan.
   E.) None of the above.

15.) Which of the following is/are true regarding Ω loops?
   A.) They generally contain 6 to 16 amino acid residues.
   B.) The end-to-end distance is less than 10 Å.
   C.) They are present at the protein surface.
   D.) They are named "omega" because they look like the "omega" symbol.
   E.) All of the above are true.

16.) Which of the following is/are true regarding fibrous proteins?
    A.) They are highly elongated molecules.
    B.) They can play a protective role in living organisms.
    C.) They can also play a connective or supportive role in living organisms.
    D.) Some fibrous proteins have motive functions in living organisms.
    E.) All of the above are true.

17.) Which of the following are the two examples of fibrous proteins that we discussed in class?
    A.) Cellulose and Chitin.
    B.) Keratin and Chitin.
    C.) Keratin and Collagen.
    D.) Collagen and Cellulose.
    E.) None of the above.

18.) What inherited skin disease did we discuss in class that can result from mutations in the genes encoding keratin 5 or keratin 14, resulting in skin blistering due to the rupturing of epidermal cells.
    A.) Epidermolysis Bullosa Simplex (EBS).
    B.) Scurvy.
    C.) Osteogenesis Imperfecta.
    D.) Tay-Sachs Disease.
    E.) None of the above.

19.) Which of the following is/are true regarding collagen?
    A.) Collagen is organized into insoluble fibers of great tensile strength.
    B.) It is the most abundant protein in vertebrates.
    C.) Collagen occurs in all multicellular animals.
    D.) Collagen is the major stress-bearing component of connective tissue.
    E.) All of the above are true.

20.) _____ is an enzyme discussed in class that catalyzes the hydroxylation of certain prolyl residues in collagen precursors, which allows for enhanced stability via increased hydrogen bonding.
    A.) Prolyl endopeptidase.
    B.) Prolylhydroxylase.
    C.) Prolase.
    D.) Prohydroxinase.
    E.) None of the above.

21.) The enzyme mentioned in question #20 above requires vitamin _____ to maintain its enzymatic activity.
    A.) A.
    B.) B.
    C.) C.
    D.) D.
    E.) None of the above.

22.) A deficiency of the vitamin mentioned in question #21 can result in which of the following as discussed in class?
    A.) Epidermolysis Bullosa Simplex (EBS).
    B.) Scurvy.
    C.) Osteogenesis Imperfecta.
    D.) Tay-Sachs Disease.
    E.) None of the above.

23.) _____ is a rare heritable disorder (also known as brittle bone disease) discussed in class resulting from the mutation of Type I collagen.
    A.) Epidermolysis Bullosa Simplex (EBS).
    B.) Scurvy.
    C.) Osteogenesis Imperfecta.
    D.) Tay-Sachs Disease.
    E.) None of the above.

24.) Which of the following is/are true regarding globular proteins?
    A.) They are compact spheroidal molecules.
    B.) Enzymes are good examples of globular proteins.
    C.) Transport and receptor proteins are also good examples of globular proteins.
    D.) Globular proteins may contain both $\alpha$ helices and $\beta$ sheets.
    E.) All of the above are true.

25.) Which of the following statements about carbonic anhydrase is incorrect?
    A.) It converts carbon dioxide to bicarbonate to maintain acid-base balance in blood and other tissues, and to help transport carbon dioxide out of tissues.
    B.) It exists as a compact spheroidal molecule.
    C.) The imidazole side chains of 3 His residues coordinate with a zinc ion in the active site.
    D.) It is a globular protein that contains both $\alpha$ helices and $\beta$ sheets.
    E.) All of the above are true.

26.) Which of the following is an example of a supersecondary structure discussed in class?
    A.) The $\beta\alpha\beta$ motif.
    B.) The $\beta$ hairpin motif.
    C.) The $\alpha\alpha$ motif.
    D.) A, B, and C are all supersecondary structures discussed in class.
    E.) None of the above.

27.) _____ stands for Nuclear Overhauser Effect Spectroscopy and is a technique discussed in class that allows for the determination of protein structure by measuring interproton distances through space.
  A.) NOESY.
  B.) CASY.
  C.) COSY.
  D.) NOBY.
  E.) None of the above.

For Questions #28 - #30, consider the NOESY spectrum in your notes (Slide #42) and the following structures:

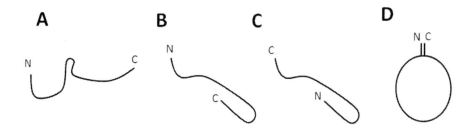

28.) Cross peak #1 from your notes (Slide #42) would most closely resemble which of the above structures?
  A.) A.
  B.) B.
  C.) C.
  D.) D.
  E.) None of the above.

29.) Cross peak #2 from your notes (Slide #42) would most closely resemble which of the above structures?
  A.) A.
  B.) B.
  C.) C.
  D.) D.
  E.) None of the above.

30.) Cross peak #3 from your notes (Slide #42) would most closely resemble which of the above structures?
  A.) A.
  B.) B.
  C.) C.
  D.) D.
  E.) None of the above.

# Chapter 5
## Answers to Additional Problem Sets (test format)

1.) B
2.) D
3.) C
4.) D
5.) A
6.) C
7.) B
8.) D
9.) A
10.) B
11.) E
12.) E
13.) E
14.) C
15.) E
16.) E
17.) C
18.) A
19.) A
20.) B
21.) C
22.) B
23.) C
24.) E
25.) E
26.) D
27.) A
28.) D
29.) B
30.) A

# Chapter 6

# *Sugars and Polysaccharides*

*Chapter 6 Summary*:

*Carbohydrates* are the most abundant class of biological molecules. The name carbohydrate means "carbon hydrate" because of its chemical composition (carbon and water). *Monosaccharides* are the basic units of carbohydrates, and are derivatives of straight-chain polyhydroxy alcohols, which are either an aldehyde (i.e. D-glucose) or a ketone (i.e. D-ribulose). They contain at least three carbon atoms, and are called simple sugars as they cannot be hydrolyzed to form simpler saccharides. Monosaccharides are classified according two major structural features, which includes both the chemical nature of the carbonyl group and number of carbon atoms. For example, the sugar is classified as an *aldose* if the carbonyl group is an aldehyde and a *ketose* if the carbonyl group is a ketone. Thus, glucose is considered to be an aldose and ribulose a ketose. With respect to the number of carbon atoms, sugars with 3, 4, 5, 6, and 7 carbons are called trioses, tetroses, pentoses, hexoses, and heptoses respectively. So, glucose is an aldohexose and ribulose is a ketopentose (see below).

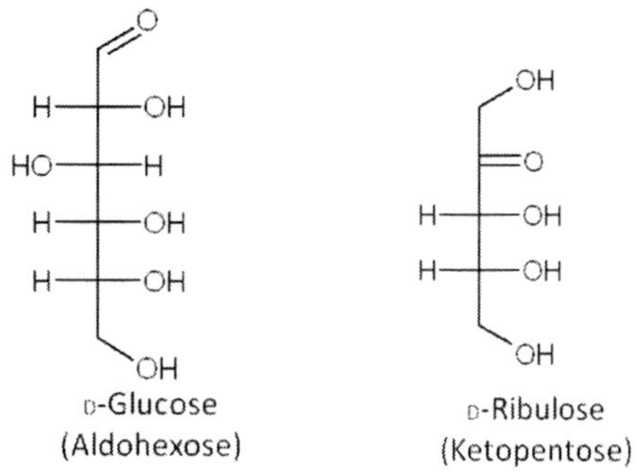

D-Glucose
(Aldohexose)

D-Ribulose
(Ketopentose)

It is also important to note that both aldoses and ketoses contain carbons with chiral centers, thus they need to be classified accordingly. The prefix D or L (according to the Fisher convention) refers to the configuration of the highest numbered asymmetric carbon farthest from the carbonyl. When comparing D and L sugars, notice that they are mirror images of each other as seen below in the Fisher projection for fructose.

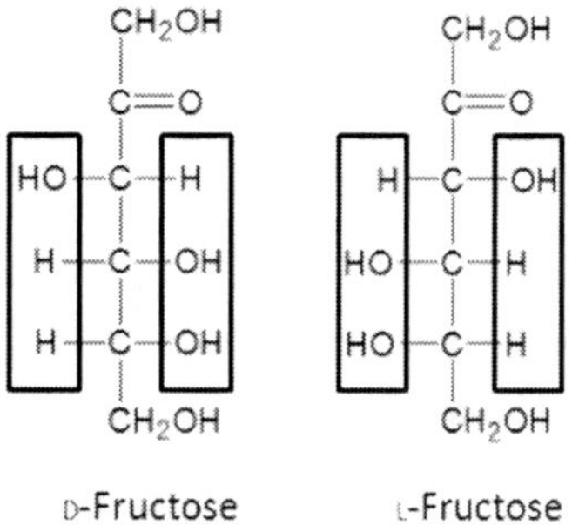

D-Fructose          L-Fructose

*Epimers* are sugars that have a different configuration about only one carbon atom (i.e. D-Glucose and D-Mannose). Cyclic hemiacetals and hemiketals can be formed from the intramolecular reaction of the alcohol group (on the furthest asymmetric carbon) with the either the aldehyde or ketone. A pyranose results if it is a six-membered ring and a furanose results if it is a five-membered ring. The result of the cyclization of a monosaccharide (commonly represented as Haworth Projection Formulas) is a pair of diastereomers known as anomers. The hemiacetal or hemiketal carbon is the anomeric carbon. If the OH group on the anomeric carbon is on the opposite side of the $CH_2OH$ group on the chiral carbon that designates the D or L configuration (i.e. C5 in glucose), then you have the α anomer. In the β form, the opposite would be true (i.e. OH group on the anomeric carbon is on the same side of the $CH_2OH$ group).

α-D-Glucopyranose      D-Glucose      β-D-Glucopyranose

It is important to remember from Organic Chemistry that a pair of diastereomers have different chemical and physical properties. Also, note that six- and five-membered rings are the focus here. That is because these ring structures are the most common due to greater stability. Seven-membered rings and greater are rarely observed, and three-and four-membered rings are unstable due to internal strain. Also, note that the structures of furanose and pyranose rings are not planar (all members of the ring are $SP^3$ hybridized).

*Derivative Forms of Monosaccharides.* A variety of chemical and enzymatic reactions produce derivatives of simple sugars. Some of the most common include sugar acids generated from redox reactions, deoxy sugars (i.e. constituents of DNA), and amino sugars which contain an amino group in place of a hydroxyl group. With respect to sugar acids, you should be familiar with the naming of three possible modifications made to glucose. This includes oxidation at C1 (gluconic acid), oxidation at C6 (glucuronic acid), and oxidation at both C1 and C6 (glucaric acid). Deoxy sugars have previously been discussed when we looked at DNA and RNA (i.e. the difference between ribonucleotides and deoxyribonucleotides). β-D-Glucosamine and β-D-Galactosamine are examples of amino sugars discussed in class that contain an amino group in place of a hydroxyl group at the C2 position.

*Oligosaccharides* consist of a few covalently linked monosaccharides, which can be associated with proteins (glycoproteins) and lipids (glycolipids) to provide structural and regulatory functions (discussed in detail later).

*Disaccharides* are the simplest oligosaccharides which contain two monosaccharides linked by a glycosidic bond. Each unit in an oligosaccharide is termed a residue. Any residue that has a free unsubstituted anomeric carbon is a reducing sugar. The common disaccharides discussed in class include Lactose, Maltose, Sucrose, Cellobiose, and Isomaltose. While it is not important to memorize these structures, you should know certain important features (i.e. identify the monosaccharides involved, be able to identify anomeric carbons and reducing/nonreducing ends). Also, be able to recognize the various types of linkages (i.e. alpha(1,4), beta(1,4), etc.).

*Polysaccharides* contain many monosaccharides linked together. They have various structural functions and also serve as nutritional reservoirs. Polysaccharides are held together by glycosidic bonds. Glycosidic bonds in polysaccharides are analogous to the peptide bond in proteins. Glycosidic bond hydrolysis is catalyzed by enzymes known as glycosidases. The nomenclature for polysaccharides is based on their composition and structure. A *homopolysaccharide* is a polysaccharide that contains only one kind of monosaccharide, while a *heteropolysaccharide* is a polysaccharide made of several monosaccharides. Polysaccharides can function as storage molecules (i.e. starch and glycogen), structural molecules (i.e. chitin and cellulose), as well as recognition molecules (i.e. cell surface polysaccharides).

*Starch* is a plant storage polysaccharide. The two forms of starch are amylose and amylopectin (both contain glucose). Most starch is 10-30% amylose and 70-90% amylopectin. Amylose has $\alpha(1\rightarrow4)$ links and one reducing end. Amylopectin contains branches (~every 24-30 residues). The branches in amylopectin are $\alpha(1\rightarrow6)$ (see below).

α-Amylose                        Amylopectin

It is also important to note that amylase is a linear polymer of several thousand glucose residues which forms an irregularly aggregating helically coiled conformation. *Phosphorylase* is an enzyme that releases glucose-1-phosphate products (energy source) from the amylose or amylopectin chains. The more branches, the more sites for phosphorylase attack, thus the branches in amylopectin provide a mechanism for quickly releasing (or storing) glucose units for (or from) metabolism.

*Glycogen* is also known as "animal starch" and constitutes up to 10% of liver mass and 1-2% of muscle mass. Glycogen is stored energy for an organism, and the only difference from amylopectin is essentially the frequency of branching. Glycogen has more branching when compared to amylopectin (i.e. $\alpha(1\rightarrow6)$ ~ every 8-14 residues).

*Cellulose and Chitin* are examples of structural polysaccharides. Chitin is the principal structural component of invertebrates (i.e. crustaceans, insects, and spiders). Cellulose is the primary structural component of rigid plant cell walls. Chitin is similar to cellulose except its C2 position is N-acetylated (see below).

Chitin                        Cellulose

Notice that the only difference between α-amylose and cellulose is the linkages. This however, makes a big difference between the two. For example, while α-amylose forms an irregularly aggregating helically coiled conformation, cellulose adopts a fully extended "ribbon" conformation.

*Glycosaminoglycans* (GAGs) are linear chains (unbranched) of alternating uronic acid and hexosamine residues. They are known as ground substances because collagen and elastin which are particularly present in connective tissue (cartilage, tendon, skin, and blood vessel walls), are found embedded in this gel-like substance. They have a slimy, mucus-like consistency resulting in high viscosity and elasticity. With respect to structure, the six glycosaminoglycans discussed in class include Hyaluronate, Chondroitin-4-sulfate, Chondroitin-6-sulfate, Keratan sulfate, Dermatan sulfate, and Heparin. Hyaluronates (consisting of up to 25,000 disaccharide units) are components of the vitreous humor of the eye and of synovial fluid. Chondroitins and keratan sulfate are found in tendons, cartilage, and other connective tissue. Dermatan sulfate is a component of the extracellular matrix of skin, and Heparin is a natural anticoagulant. GAGs are also constituents of proteoglycans (discussed later).

*Glycoproteins* are proteins that contain oligosaccharide chains covalently attached to their polypeptide side-chains (can be O-linked and N-linked). Most Eukaryotic proteins exist as glycoproteins. Glycoproteins vary in their carbohydrate content (<1% – >90%). Known glycoproteins serve as membrane receptors (proteoglycans such as syndecan) immunoglobins (antibodies), and structural proteins (peptidoglycan found in bacterial cell walls). The *N-linked* oligosaccharides discussed in class that you should be familiar with consists of an N-acetylglucosamine (NAG) sugar β-linked to the amide nitrogen of an asparagine residue. The *O-linked* oligosaccharides discussed in class that you should also be familiar with consists of an N-acetylgalactosamine sugar α-linked to the hydroxyl group of either serine or threonine.

*Proteoglycans* are a diverse group of macromolecules. They are found in the extracellular matrix between cells ("fillers"). They can also be cell-membrane receptors (i.e. Syndecan). They consist of a core protein, to which at least one glycosaminoglycan chain is covalently linked. Keratan sulfate and Chondroitin sulfate are the most common glycosaminoglycans found linked to the core protein.

*Peptidoglycan* is another example of a glycoprotein. It serves as a structural role in the bacterial cell wall, resulting in significant strength. Penicillin binds to and inactivates enzymes that function to cross-link peptidoglycan strands of bacterial cell walls. Most bacteria that are resistant to penicillin secrete penicillinase, which inactivates penicillin by cleaving the amide bond of its β-lactam ring.

*Chapter 6*
*Lecture Series*

Slide #1 – Introduction
*Notes*

Slide #2 – Carbohydrates are the most abundant class of biological molecules. The name carbohydrate literally means "carbon and hydrate" because its chemical composition is basically carbon and water. _____ are the basic units of carbohydrates, which are synthesized through _____, and some are products of photosynthesis. If you covalently link several monosaccharides, you get oligosaccharides, which are commonly associated with proteins (called glycoproteins) and lipids (called glycolipids) for various structural and regulatory functions.
*Notes*

Slide #3 – Polysaccharides consist of many covalently linked monosaccharide units. They can have molecular masses ranging well into the millions of daltons. They also have various structural functions (i.e. _____ in plants) and also serve as nutritional reservoirs (i.e. starch in plants and _____ in animals).
*Notes*

Slide #4 – _____ are aldehyde (i.e. D-Glucose) or ketone (i.e. D-Ribulose) derivatives of straight-chain polyhydroxy alcohols. They contain at least _____ carbon atoms. They are also called simple sugars and cannot be hydrolyzed to form simpler saccharides. Can you label the following?
*Notes*

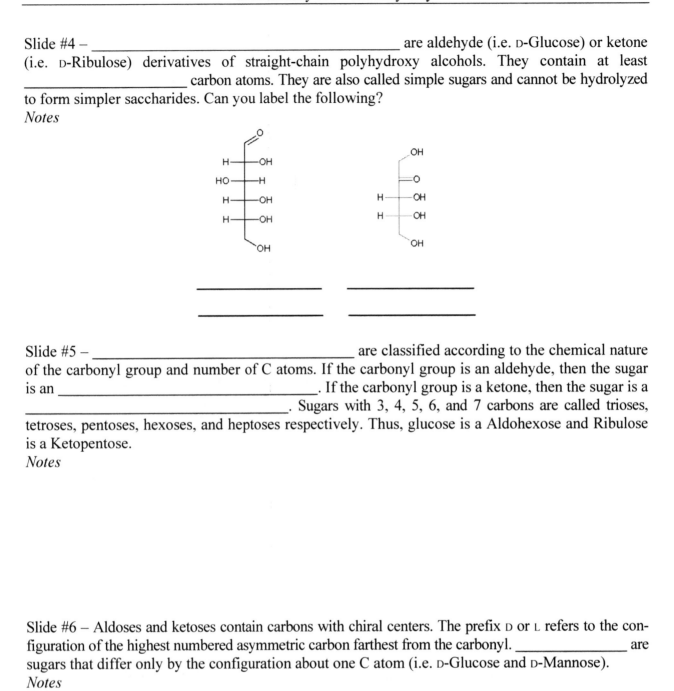

_____        _____

_____        _____

Slide #5 – _____ are classified according to the chemical nature of the carbonyl group and number of C atoms. If the carbonyl group is an aldehyde, then the sugar is an _____. If the carbonyl group is a ketone, then the sugar is a _____. Sugars with 3, 4, 5, 6, and 7 carbons are called trioses, tetroses, pentoses, hexoses, and heptoses respectively. Thus, glucose is a Aldohexose and Ribulose is a Ketopentose.
*Notes*

Slide #6 – Aldoses and ketoses contain carbons with chiral centers. The prefix D or L refers to the configuration of the highest numbered asymmetric carbon farthest from the carbonyl. _____ are sugars that differ only by the configuration about one C atom (i.e. D-Glucose and D-Mannose).
*Notes*

Slide #7 – Fructose. Can you tell which one is D and L?
*Notes*

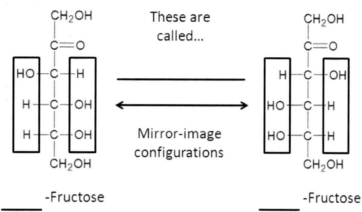

Slide #8 – The D-Aldoses. While it is not necessary to memorize all of these structures, you should be familiar with common sugars (i.e. Glucose and Galactose etc.).
*Notes*

Slide #9 – The D-Ketoses. While it is not necessary to memorize all of these structures, you should be familiar with common sugars (i.e. Fructose).
*Notes*

Slide #10 – Hemiacetals and Hemiketals result from the reaction of an alcohol with either an
_____ or a _____ respectively. Can you draw a
Hemiacetal and a Hemiketal?
*Notes*

Slide #11 –
*Notes*

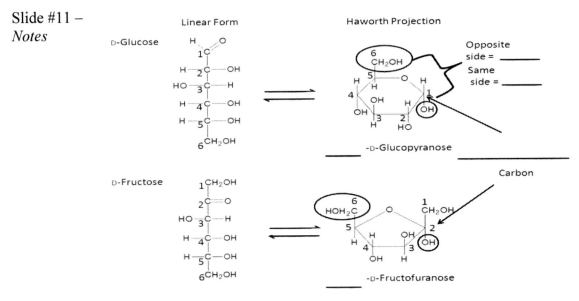

Slide #12 – As with any pair of _____, α-D-Glucopyranose
and β-D-Glucopyranose have very different chemical and physical properties.
*Notes*

142

Slide #13 – Monosaccharide Structures. Simple sugars can cyclize in two ways, forming either furanose or pyranose structures.
*Notes*

Slide #14 – Monosaccharide Structures. Simple sugars can cyclize in two ways, forming either furanose or pyranose structures.
*Notes*

Slide #15 – Sugar Confirmations. _____- and _____-membered rings are the most common due to greater stability. Seven-membered rings and greater are rarely observed. Three- and four-membered rings are unstable due to internal strain. The furanose and pyranose rings are NOT planar (all members of the ring are _____hydridized).
*Notes*

Slide #16 – The pyranose ring can assume the boat and chair conformations. In the chair conformation, ring substituents fall into different geometrical classes, which are referred to as the _____ or _____ position. Hydroxyls that are in the _____ position are more reactive than in the axial position. Can you draw both conformations?

*Notes*

Slide #17 – A variety of chemical and enzymatic reactions produce derivatives of the simple sugars Some of the most common are:

-_____ acids that are generated from Redox reactions.

-_____ sugars (i.e. constituents of DNA, etc.).

-_____ sugars which contain an amino group in place of a hydroxyl group.

*Notes*

Slide #18 – Sugar Acids.
*Notes*

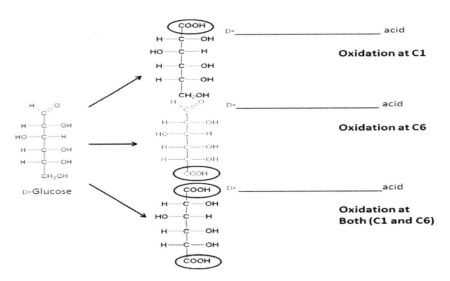

Slide #19 – Deoxy Sugars.
*Notes*

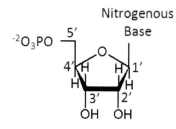 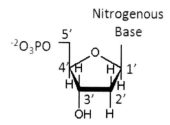

_____          _____

Slide #20 – Amino Sugars. Amino sugars - contain an amino group in place of a hydroxyl group at the _____ position.
*Notes*

_____          _____

Slide #21 – _____ are the simplest oligosaccharides. Here we have two monosaccharides linked by a _____ bond. Each unit in an oligosaccharide is termed a residue, and any residue that has a free unsubstituted anomeric carbon is a reducing sugar.
*Notes*

Slide #22 –
*Notes*

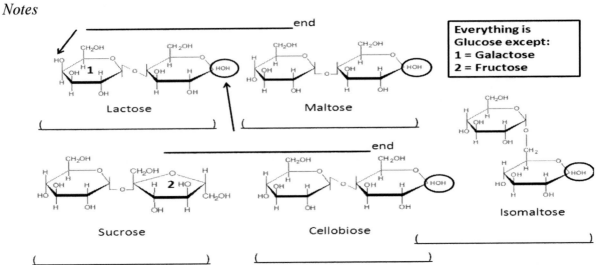

Slide #23 – While it's not important to memorize these structures, you should know important features. Be able to identify anomeric carbons and reducing and nonreducing ends. ***Note that sucrose is NOT a _____ sugar. Also, note carefully the nomenclature of the links. Be able to recognize alpha(1,4), beta(1,4) linkages, etc.
*Notes*

Slide #24 – Oligosaccharides occur widely as components of antibiotics (derived from various sources). _____ is an oligosaccharide produced by *Stretomyces griseus*.
*Notes*

Slide #25 – _____ are longer chains of oligosaccharides. Polysaccharides are held together by glycosidic bonds. Glycosidic bonds in polysaccharides are analogous to the peptide bond in proteins. Glycosidic bond hydrolysis is catalyzed by enzymes known as _____. The nomenclature for polysaccharides is based on their composition and structure.
*Notes*

Slide #26 – A _____ is a polysaccharide that contains only one kind of monosaccharide, and a _____ is a polysaccharide made of several monosaccharides. Polysaccharides functions as:

      a.) _____ molecules (i.e. Starch and glycogen ).
      b.) _____ molecules (i.e. Chitin and cellulose).
      c.) _____ molecules (i.e. Cell surface polysaccharides).
*Notes*

Slide #27 – _____ is a plant storage polysaccharide. The two forms of it include _____and _____. Most starch is 10-30% amylose and 70-90% amylopectin. Amylose has alpha(1,4) links and one reducing end. Branches in amylopectin occur every 24-30 residues. The branches in amylopectin are α_____.
*Notes*

Slide #28 – _____ is a linear polymer of several thousand glucose residues linked by α(1, 4) linkages. It forms an irregularly aggregating helically _____ conformation.

*Notes*

CH₂OH        CH₂OH

Glucose        Glucose        *n*

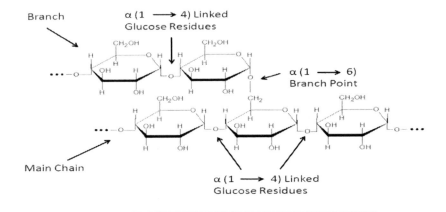

Slide #29 – _____ consists mainly of α(1,4)–linked glucose residues, but is a branched molecule with a α(1,6) branch every _____ glucose residues.

*Notes*

Branch

α(1 ⟶ 4) Linked
Glucose Residues

α(1 ⟶ 6)
Branch Point

Main Chain

α(1 ⟶ 4) Linked
Glucose Residues

Slide #30 – The _____ reaction releases glucose-1-P products (energy source) from the amylose or amylopectin chains. The more branches, the more sites for _____ attack. Branches in amylopectin provide a mechanism for quickly releasing (or storing) glucose units for (or from) metabolism.

*Notes*

Slide #31 – Let's take a look at this reaction.
*Notes*

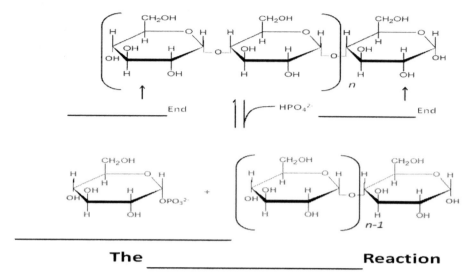

Slide #32 – _____ is "Animal Starch" and is the storage device for energy in animals. It constitutes up to 10% of liver mass and 1-2% of muscle mass. The only difference from amylopectin is the frequency of branching, which is approximately every _____ residues (alpha (1,6) branches).
*Notes*

Slide #33 – _____ have a small but significant difference from starch and glycogen. If you change the main linkages between glucose from alpha(1,4) to _____, you get this new family of polysaccharides. These also have branches which can be (1,2), (1,3), or (1,4). Cross-linked dextrans are used as "_____" gels in column chromatography. These gels, used to separate biomolecules on the basis of size, are up to 98% water!
*Notes*

Slide #34 – _____ and _____ are examples of structural polysaccharides. _____ is the principal structural component of invertebrates (i.e. crustaceans, insects, and spiders), and is similar to cellulose except its C2 is _____. Finally, _____ is the primary structural component of rigid plant cell walls.
*Notes*

Slide #35 – Structures of Chitin and Cellulose.
*Notes*

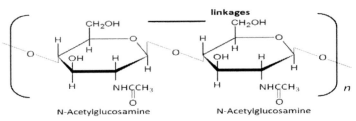

**Name:**

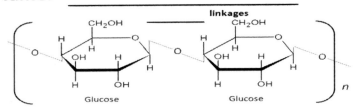

**Name:** _____

Slide #36 – _____ is the most abundant natural polymer in the world. It is found in the wood and bark of trees. It is found in the _____ of nearly all plants. Cotton is almost pure cellulose. It is an _____ of α-amylose (with very different physical properties).
*Notes*

Slide #37 – Comparison between α-Amylose and Cellulose. Note, both Cellulose and Chitin adopt a fully extended "_____" conformation.
*Notes*

**Name:**

_____

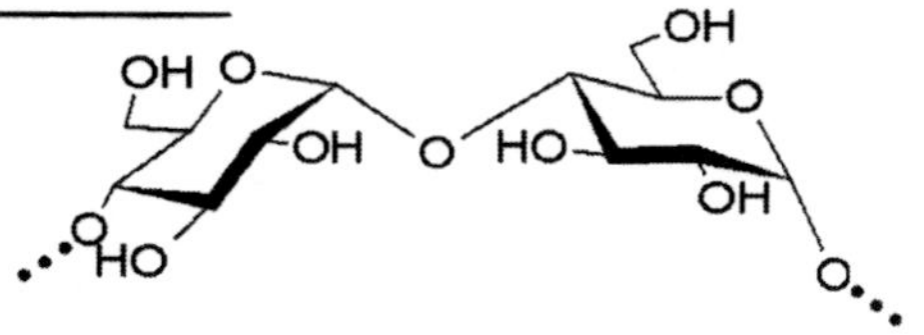

**Name:**

_____

_____ **D-Glucose Units**

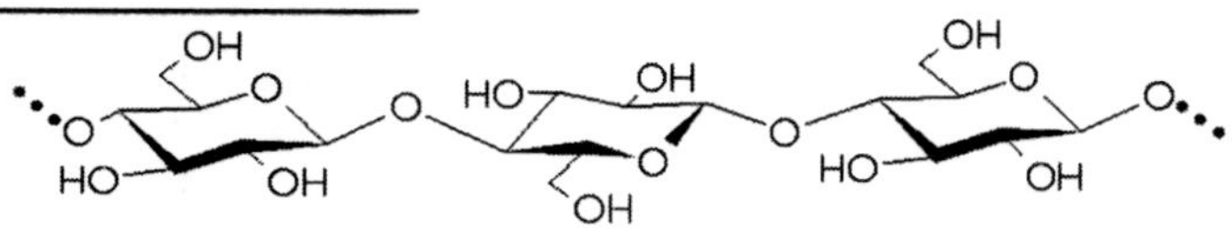

_____ **D-Glucose Units**

Slide #38 – Cellulose fibers have exceptional strength due to its highly cohesive, hydrogen bonded structure. Note, that both Cellulose and Chitin form _____ bonds.
*Notes*

Slide #39 – Ruminants (i.e. Giraffes, cattle, deer, camels, buffalos) are animals that are able to metabolize cellulose. They have a bacterial cellulase in their rumen. Cellulase is considered a _____. A rumen is a large first compartment in their stomach.
*Notes*

Slide #40 – Other Structural Polysaccharides. _____ is a component of agar obtained from marine red algae, that is composed of galactose polymers (used in laboratories to separate biomolecules). _____ are repeating disaccharides with amino sugars and negative charges.
*Notes*

Slide #41 – _____ are linear chains (unbranched) of alternating _____ acid and hexosamine residues. They are known as ground substance because collagen and elastin, which are particularly present in connective tissue (cartilage, tendon, skin, and blood vessel walls), are found embedded in this gel-like substance. They have a slimy, mucus-like consistency resulting in high viscosity and elasticity.
*Notes*

Slide #42 –
*Notes*

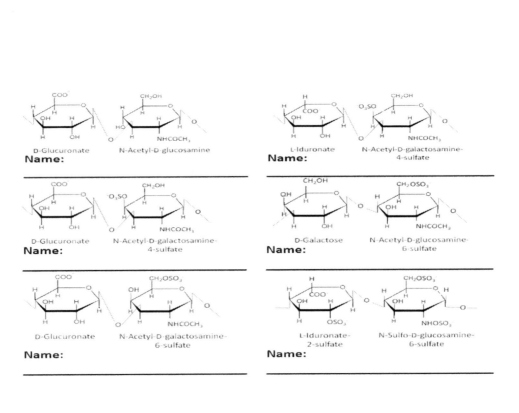

Slide #43 – _____ is a natural anticoagulant. Hyaluronates (consisting of up to 25,000 disaccharide units) are components of the vitreous humor of the eye and of synovial fluid, the lubricant fluid of the body's joints. Chondroitins and keratan sulfate are found in tendons, cartilage, and other connective tissue. Dermatan sulfate is a component of the extracellular matrix of skin. Glycosaminoglycans are constituents of _____ (discussed later).

*Notes*

Slide #44 – _____ are proteins that contain oligosaccharide chains covalently attached to their polypeptide side-chains (can be O-linked and N-linked). Most _____ proteins exist as glycoproteins. Glycoproteins vary in their carbohydrate content (<1% – >90%). Known glycoproteins serve as membrane receptors (proteoglycans such as _____), immunoglobins (antibodies), and structural proteins (peptidoglycan found in bacterial cell walls).

*Notes*

Slide #45 – _____ oligosaccharides contain an _____(NAG) that is _____-linked to the amide nitrogen of an _____ residue.

*Notes*

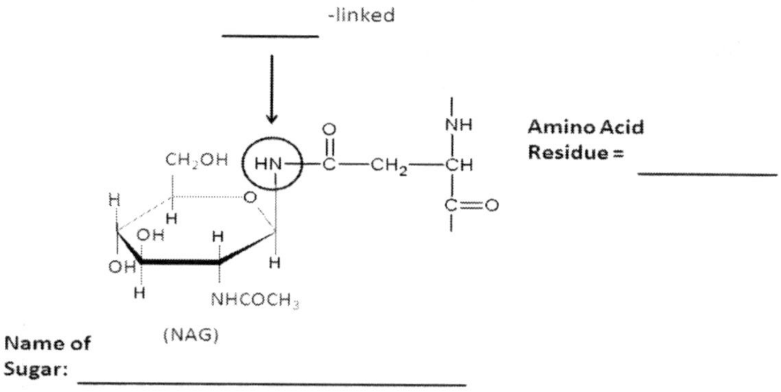

Slide #46 – _____ oligosaccharides contain
an _____ that is _____-linked to the hydroxyl group of either
_____ or _____.

*Notes*

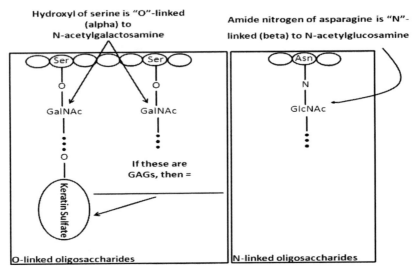

Slide #47 – O-linked and N-linked Oligosaccharides.
*Notes*

Slide #48 – _____ are a diverse group of macromolecules. They are found
in the extracellular matrix between cells ("filler"). They can also be cell-membrane receptors
(Syndecan). They consist of a core protein, to which at least one _____ chain
is covalently linked. Keratan sulfate and chondroitin sulfate are the most common glycosaminoglycans found linked to the core protein.
*Notes*

Slide #49 – _____ are single transmembrane domain proteins. They are coreceptors that contain heparan sulfate and chondroitin sulfate chains. They can bind a variety of ligands (i.e. growth factors).
*Notes*

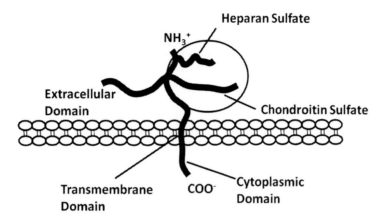

Slide #50 – A look at how this relates to previous research.
*Notes*

Slide #51 – A look at how this relates to previous research.
*Notes*

Slide #52 – _____ binds to and inactivates enzymes that function to cross-link peptidoglycan strands of bacterial cell walls. Most bacteria that are resistant to this secrete _____, which inactivates penicillin by cleaving the amide bond of its _____ ring.

*Notes*

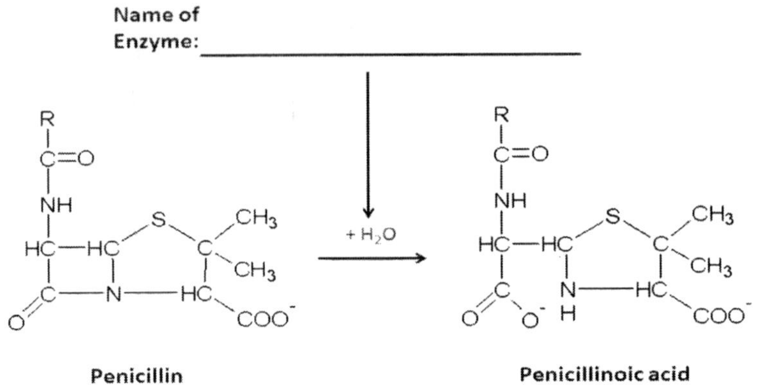

Name of Enzyme: _____

Penicillin          + $H_2O$          Penicillinoic acid

Slide #53 – Recommended Problem Sets (from the textbook).

*Notes*

## *Chapter 6*
## *Additional Problem Sets (test format)*

1.) Which of the following terms can be correctly used when describing carbohydrates?
   A.) Monosaccharides.
   B.) Disaccharides.
   C.) Polysaccharides.
   D.) Oligosaccharides.
   E.) All of the above.

2.) Oxidation at C1 of D-Glucose results in a sugar acid. The correct name for this molecule is _____.
   A.) Glucaric acid.
   B.) Glucuronic acid.
   C.) Gluconic acid.
   D.) Glucose.
   E.) None of the above.

3.) Oxidation at C6 of D-Glucose results in a sugar acid. The correct name for this molecule is _____.
   A.) Glucaric acid.
   B.) Glucuronic acid.
   C.) Gluconic acid.
   D.) Glucose.
   E.) None of the above.

4.) Oxidation at both C1 and C6 of D-Glucose results in a sugar acid. The correct name for this molecule is _____.
   A.) Glucaric acid.
   B.) Glucuronic acid.
   C.) Gluconic acid.
   D.) Glucose.
   E.) None of the above.

5.) Which of the following is/are true regarding disaccharides?
   A.) They are the simplest oligosaccharides.
   B.) They are composed of two monosaccharides.
   C.) The two monosaccharides are linked by a glycosidic bond.
   D.) The glycosidic bond can be either of the α or β configuration.
   E.) All of the above.

For Questions #6 - #10, consider the following two disaccharide structures:

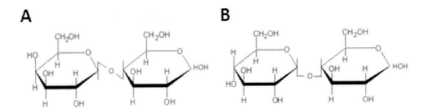

6.) Which of the above structures contain an α-linkage?
  A.) Structure "A" only.
  B.) Structure "B" only.
  C.) Both structure "A" and "B" contain an α-linkage.
  D.) Neither structure contains an α-linkage.
  E.) This cannot be determined from the structures as drawn here.

7.) Which of the above structures contain a β-linkage?
  A.) Structure "A" only.
  B.) Structure "B" only.
  C.) Both structure "A" and "B" contain a β-linkage.
  D.) Neither structure contains a β-linkage.
  E.) This cannot be determined from the structures as drawn here.

8.) Which of the above structures contain a galactose residue?
  A.) Structure "A" only.
  B.) Structure "B" only.
  C.) Both structure "A" and "B" contain a galactose residue.
  D.) Neither structure contains a galactose residue.
  E.) This cannot be determined from the structures as drawn here.

9.) Another name for lactose (structure "A") is _____.
  A.) Glucose-α-1,4-glucose.
  B.) Galactose-α-1,4-glucose.
  C.) Glucose-β-1,4-glucose.
  D.) Galactose-β-1,4-glucose.
  E.) None of the above.

10.) Another name for maltose (structure "B") is _____.
  A.) Glucose-α-1,4-glucose.
  B.) Galactose-α-1,4-glucose.
  C.) Glucose-β-1,4-glucose.
  D.) Galactose-β-1,4-glucose.
  E.) None of the above.

11.) Which of the following statements about polysaccharides is <u>incorrect</u>?
    A.) They are longer chains of oligosaccharides.
    B.) They are held together by glycosidic bonds.
    C.) Glycosidic bonds in polysaccharides are analogous to the peptide bond in proteins.
    D.) Glycosidic bond hydrolysis is catalyzed by enzymes known as glycosidases.
    E.) All of the above statements are true.

12.) Which of the following polysaccharides is <u>not</u> an example of a storage molecule in either plants or animals?
    A.) Glycogen.
    B.) Cellulose.
    C.) Amylose.
    D.) Amylopectin.
    E.) All of the above are storage molecules either in plants or animals.

13.) Which of the following statements regarding starch is/are true?
    A.) Starch is a plant storage polysaccharide.
    B.) The two forms of starch are amylose and amylopectin.
    C.) Both amylose and amylopectin contain glucose.
    D.) Most starch is 10-30% amylose and 70-90% amylopectin.
    E.) All of the above statements are true.

14.) In plant cells, starch is hydrolyzed by _____ to release _____ and a starch molecule with one less sugar residue.
    A.) Starch phosphorylase; glucose-1-phosphate.
    B.) Salivary α-amylase; maltose 1-phosphate.
    C.) Starch phosphorylase; galactose-1-phosphate.
    D.) Starch phosphorylase; fructose-1-phosphate.
    E.) None of the above.

15.) Which of the following statements regarding glycogen is/are true?
    A.) It is sometimes referred to as "Animal Starch".
    B.) It allows animals to store glucose.
    C.) Glycogen is stored energy for an organism.
    D.) Essentially the only difference from amylopectin is the frequency of branching (more branching in glycogen).
    E.) All of the above statements are true.

16.) _____ is a glycosaminoglycan and is a natural anticoagulant.
    A.) Heparin.
    B.) Hyaluronate.
    C.) Dermatin Sulfate.
    D.) Keratan Sulfate.
    E.) None of the above.

17.) _____ is a glycosaminoglycan mostly found in skin, and is therefore named as such due to its prevalence in the skin.
   A.) Heparin.
   B.) Hyaluronate.
   C.) Dermatan Sulfate.
   D.) Keratan Sulfate.
   E.) None of the above.

18.) _____ is a glycosaminoglycan and is known as the lubricant fluid of the body (synovial fluid). It is also an important component of the vitreous humor of the eye.
   A.) Heparin.
   B.) Hyaluronate.
   C.) Dermatan Sulfate.
   D.) Keratan Sulfate.
   E.) None of the above.

19.) Glycosaminoglycans are unbranched polysaccharides consisting of alternating residues of uronic acid and hexosamine. Which of the following is NOT a glycosaminoglycan?
   A.) Heparin
   B.) Hyaluronate.
   C.) Dermatan Sulfate.
   D.) Keratan Sulfate.
   E.) All of the above are glycosaminoglycans.

20.) Which of the following statements regarding glycoproteins is/are true?
   A.) Glycoproteins are proteins that contain oligosaccharide chains.
   B.) The oligosaccharide chains are covalently attached to polypeptide side-chains.
   C.) Attachment of oligosaccharide chains to polypeptide side-chains can be either O-linked or N-linked.
   D.) Most Eukaryotic proteins exist as glycoproteins.
   E.) All of the above statements are true.

For Questions #21 - #28, consider the following two glycoprotein linkages:

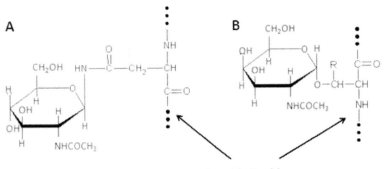

Peptide Backbone

21.) Which of the above structures contain an N-acetylglucosamine residue?
  A.) Structure "A" only.
  B.) Structure "B" only.
  C.) Both structure "A" and "B" contain an N-acetylglucosamine residue.
  D.) Neither structure contains an N-acetylglucosamine residue.
  E.) This cannot be determined from the structures as drawn here.

22.) Which of the above structures contain an N-acetylgalactosamine residue?
  A.) Structure "A" only.
  B.) Structure "B" only.
  C.) Both structure "A" and "B" contain an N-acetylgalactosamine residue.
  D.) Neither structure contains an N-acetylgalactosamine residue.
  E.) This cannot be determined from the structures as drawn here.

23.) Which of the above structures contain a β-linkage?
  A.) Structure "A" only.
  B.) Structure "B" only.
  C.) Both structure "A" and "B" contain a β-linkage.
  D.) Neither structure contains a β-linkage.
  E.) This cannot be determined from the structures as drawn here.

24.) Which of the above structures contain an α-linkage?
  A.) Structure "A" only.
  B.) Structure "B" only.
  C.) Both structure "A" and "B" contain an α-linkage.
  D.) Neither structure contains an α-linkage.
  E.) This cannot be determined from the structures as drawn here.

25.) Which of the above structures can be considered an N-linked oligosaccharide?
  A.) Structure "A" only.
  B.) Structure "B" only.
  C.) Both structure "A" and "B" can be considered an N-linked oligosaccharide.
  D.) Neither structure can be considered an N-linked oligosaccharide.
  E.) This cannot be determined from the structures as drawn here.

26.) Which of the above structures can be considered an O-linked oligosaccharide?
  A.) Structure "A" only.
  B.) Structure "B" only.
  C.) Both structure "A" and "B" can be considered an O-linked oligosaccharide.
  D.) Neither structure can be considered an O-linked oligosaccharide.
  E.) This cannot be determined from the structures as drawn here.

27.) Which of the above structures contain an asparagine residue?
    A.) Structure "A" only.
    B.) Structure "B" only.
    C.) Both structure "A" and "B" contain an asparagine residue.
    D.) Neither structure contains an asparagine residue.
    E.) This cannot be determined from the structures as drawn here.

28.) Which of the above structures contain either a serine or threonine residue (depending on the identity of "R")?
    A.) Structure "A" only.
    B.) Structure "B" only.
    C.) Both structure "A" and "B" contain either a serine or threonine residue.
    D.) Neither structure contains contain either a serine or threonine residue.
    E.) This cannot be determined from the structures as drawn here.

29.) Which of the following statements regarding proteoglycans is/are true?
    A.) Proteoglycans are a diverse group of macromolecules.
    B.) They are found in the extracellular matrix between cells (i.e. "fillers").
    C.) They can also be cell-membrane receptors.
    D.) They consist of a core protein, to which at least one glycosaminoglycan chain is covalently linked.
    E.) All of the above statements are true.

30.) Look at all of the structures discussed in this chapter and be sure you can recognize important features. Be able to identify different sugars (i.e. glucose versus galactose), modified sugars (i.e. N-acetylglucosamine etc.), anomeric carbons and reducing and nonreducing ends. Be able to recognize alpha(1,4), beta(1,4) linkages, etc., in all structures.

# Chapter 6
## *Answers to Additional Problem Sets (test format)*

1.) E
2.) C
3.) B
4.) A
5.) E
6.) B
7.) A
8.) A
9.) D
10.) A
11.) E
12.) B
13.) E
14.) A
15.) E
16.) A
17.) C
18.) B
19.) E
20.) E
21.) A
22.) B
23.) A
24.) B
25.) A
26.) B
27.) A
28.) B
29.) E
30.) Look at all structures discussed in this chapter in your textbook and practice!

# Chapter 7

## *Lipids and Membranes*

*Chapter 7 Summary*:

*Biological membranes* are assemblies of lipids and proteins with a small amount of carbohydrates. They regulate the composition of the intracellular medium by controlling the flow of nutrients, waste products, ions etc. in and out of the cell. They contain embedded pumps and gates that serve to transport substances against an electrochemical gradient. The word "lipids" is derived from the Greek word for fat, or "lipos". Lipids are soluble in organic solvents (i.e. chloroform and methanol), and are sparingly soluble (if at all) in water. They can be easily separated from other biological materials by extraction into organic solvents. Fats, oils, certain vitamins and hormones, and most nonprotein membrane components are lipids. *Fatty acids* are carboxylic acids with long-chain hydrocarbon side groups. They are rarely free in nature (occur as esterified components of various lipids). The $C_{16}$ and $C_{18}$ species of palmitic, oleic, linoleic, and stearic acids are the most common in higher plants and animals (<14 or >20 carbon atoms uncommon). Some common biological fatty acids you should know are as follows:

## Common Biological Fatty Acids

| Symbol | Common Name | Structure |
|--------|-------------|-----------|
| *Saturated Fatty Acids* | | |
| 12:0 | Lauric Acid | $CH_3(CH_2)_{10}COOH$ |
| 14:0 | Myristic Acid | $CH_3(CH_2)_{12}COOH$ |
| 16:0 | Palmitic Acid | $CH_3(CH_2)_{14}COOH$ |
| 18:0 | Stearic Acid | $CH_3(CH_2)_{16}COOH$ |

| Symbol | Common Name | Structure |
|--------|-------------|-----------|
| *Unsaturated Fatty Acids* | | |
| 16:1 | Palmitoleic Acid | $CH_3(CH_2)_5CH = CH(CH_2)_7COOH$ |
| 18:1 | Oleic Acid | $CH_3(CH_2)_7CH = CH(CH_2)_7COOH$ |
| 18:2 | Linoleic Acid | $CH_3(CH_2)_4(CH = CHCH_2)_2(CH_2)_6COOH$ |

*Saturated* chains pack tightly and form more rigid organized aggregates, while *unsaturated* chains bend and pack in a less ordered way with greater potential for motion. Notice that in unsaturated fatty acids, the first double bond generally appears between $C_9$ and $C_{10}$. Triple bonds rarely occur in fatty acids. *Triacylglycerols* are nonpolar, water insoluble substances that consist of fatty acid triesters of glycerol. Fats and oils in plants and animals consist of mixtures of triacylglycerols. They function as energy reservoirs in animals and are the most abundant class of lipids. They are NOT components of membranes. *Simple triacylglycerols* contain only one type of fatty acid (i.e. tristearoylglycerol), while *mixed triacylglycerols* contain two or three different types of fatty acid residues. They are named according to the placement on the glycerol moiety, and are much more common in nature than simple triacylglycerols. *Glycerphospholipids* are the major lipid components of biological membranes, and consist of sn-glycerol-3-phosphate esterified at its $C_1$ and $C_2$ positions to fatty acids. They are therefore amphiphilic molecules. Saturated $C_{16}$ and $C_{18}$ fatty acids usually occur at the $C_1$ position, while unsaturated fatty acids ($C_{16}$-$C_{20}$) usually occur at the $C_2$ position. Glycerphospholipids are named according to the identities of the fatty acids (see table above) as well as the phosphate head group. As for the phosphate head groups, you should be familiar with phosphatidic acid, phosphatidylethanolamine, and phosphatidylcholine (see below):

## Common Phospholipid Head Groups

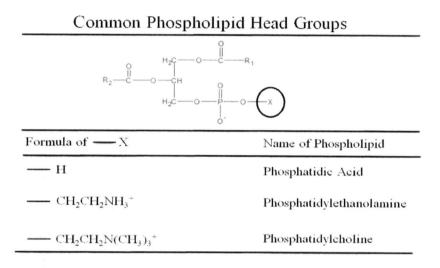

| Formula of ——X | Name of Phospholipid |
|---|---|
| —— H | Phosphatidic Acid |
| —— $CH_2CH_2NH_3^+$ | Phosphatidylethanolamine |
| —— $CH_2CH_2N(CH_3)_3^+$ | Phosphatidylcholine |

*Ether glycerophospholipids* possess an ether linkage at the $C_1$ position of glycerol (instead of an ester linkage). *Plasmalogens* are ether glycerophospholipids in which the alkyl chain is unsaturated. *Platelet activating factor* (PAF) is an ether glycerophospholipid in which the alkyl chain is saturated. *Sphingolipids* represent another class of lipids found frequently in biological membranes. They are derivatives of sphingosine ($C_{18}$ amino alcohols). The sphingosine moiety forms the backbone of these lipids rather than glycerol. A fatty acid joined to sphingosine via an amide linkage forms a ceramide. *Sphingomyelins* are the most common sphingolipid. They contain a ceramide with either a phospho-choline or a phosphoethanolamine head group. They can therefore be classified as a sphingophos-pholipid. Sphingomyelins can be found in myelin sheaths that surrounds and insulates many nerve cell axons. *Cerebrosides* are considered the simplest sphingoglycolipid. They contain a ceramide with one or more sugars in a beta-glycosidic linkage at the 1-hydroxyl group. *Galactocerebrosides* (β-D-galactose head group) are prevalent in neuronal cell membranes of the brain, while *glucocerebrosides* (β-D-glucose residue) are found in cell membranes of various other tissues. *Gangliosides* are the most complex group of the sphingoglycolipids (ceramide oligosaccharides). They all have among the sugar groups at least one sialic acid residue. There are over 60 ganglioside structures that have been identified. They act as specific receptors for certain pituitary glycoprotein hormones that regulate physiologically important functions. Disorders of ganglioside breakdown are responsible for several hereditary diseases (i.e. Tay-Sachs). *Steroids* are derivatives of a molecule containing four fused saturated rings. Cholesterol is a sterol and is the most common steroid in animals and the precursor for all other steroids in animals. Examples include cortisol (provides control of carbohydrate, protein, and lipid metabolism), testosterone (primary male sex steroid hormone), estradiol (primary female sex steroid hormone), and progesterone (precursor of testosterone and estradiol).

*Biological Signals.* Glycerophospholipids and sphingolipids play important roles as chemical signals in and on cells. Lipid signals act locally, either within the cell where they are made or on nearby cells (unlike steroid hormones). These signals typically initiate a cascade of reactions with multiple effects. The lifetimes of these signals are usually very short, and the creation/breakdown of lipid signals is carefully regulated and timed. The action of phospholipases can produce such signals. For example, phospholipases $A_1$ and $A_2$ cleave fatty acids from glycerophospholipids, producing lysophospholipids. Phospholipases C and D hydrolyze on either side of the phosphate moiety in the polar head group. Another example of lipid metabolites acting as biological signals is sphingosine-1-phosphate (S1P). Excreted S1P binds to different receptor proteins and provokes many different cell and tissue effects.

*Lipid Aggregates.* The two lipid aggregates discussed here are micelles and liposomes. *Micelles* are formed by amphiphilic molecules (i.e. soaps and detergents). The molecular arrangement eliminates unfavorable interactions between water and the hydrophobic tails of the amphiphiles. Dilute aqueous solutions of amphiphiles do not form micelles. The concentration of amphiphile molecules required in order to form micelles is referred to as the Critical Micelle Concentration (CMC). *Liposomes* were first described in the 1960s by Alec Bangham. They are formed from sonication (agitation with ultrasonic vibrations) of multilamellar vesicles (formed from suspending phospholipids in water) to yield unilamellar vesicles. Therefore, they have a phospholipid bilayer. Both micelles and liposomes have been used as drug delivery vehicles in cancer therapy. Micelles can accommodate hydrophobic drugs in the internal hydrophobic core. Liposomes can accommodate both hydrophobic as well as hydrophilic chemotherapeutics either in the hydrophobic bilayer or the internal aqueous core respectively.

*Lipid Bilayers.* Many early studies used liposomes as models to understand the cell bilayer. Using liposomes, it was determined early on that lipid bilayers are extremely impermeable to ionic and polar substances. However, water (despite its polarity) is able to penetrate the bilayer due to its small size. *Transverse diffusion* (or flip-flop) is a rare event involving the transfer of lipid molecules across the bilayer to the other side. *Lateral diffusion* is very common because lipids are highly mobile in the plane of the bilayer. The *transition temperature* is the temperature at which phospholipids undergo a phase change (order-disorder transition). Above the transition temperature, the lipids are very mobile and "fluid-like". However, below the transition temperature, the lipids are less mobile and become a "gel-like" solid. Transition temperatures are unique to each phospholipid as they are dependant in part on the fatty acid chain length and degree of saturation.

*Biological Membranes.* Biological membranes are composed of proteins associated with a lipid matrix. Specific proteins occur only in particular membranes. *Integral* (intrinsic) proteins are tightly bound to membranes via hydrophobic forces, while *peripheral* (extrinsic) proteins associate with the membrane by binding at its surface to lipid head groups (and/or integral proteins) through electrostatic interactions or hydrogen bonding.

*Vitamins.* Vitamins are compounds that cannot be synthesized in sufficient quantities by an organism. Therefore, they have to be obtained from the diet. Vitamins are classified by their biological and chemical activity, not their structure. Vitamins can be either water soluble or fat soluble. The fat (lipid) soluble vitamins are A, D, E, and K.

## Chapter 7
### Lecture Series

Slide #1 – Introduction
*Notes*

Slide #2 – Membranes. Biological membranes are assemblies of lipids and proteins (small amounts of _____. They Regulate the composition of the intracellular medium by controlling the flow of nutrients, waste products, ions etc. in and out of the cell. They contain embedded pumps and gates which can be used to transport substances against an electro-chemical gradient.
*Notes*

Slide #3 – Lipid Classification. The word lipids is derived from the Greek word for fat "_____". They are soluble in organic solvents (i.e. chloroform and methanol), and sparingly soluble (if at all) in water. They can be easily separated from other biological materials by extraction into organic solvents. Fats, oils, certain vitamins and hormones, and most nonprotein membrane components are _____.
*Notes*

Slide #4 – Fatty acids are carboxylic acids with long-chain hydrocarbon side groups. They are rarely free in nature, rather they occur as _____ components of various lipids. The $C_{16}$ and $C_{18}$ species of palmitic, oleic, linoleic, and stearic acids are the most common in higher plants and animals (<14 or >20 carbon atoms uncommon). About half of the of the fatty acid residues of plant animal lipids are _____.
*Notes*

Slide #5 – Common Biological Fatty Acids.
*Notes*

# Common Biological Fatty Acids

| Symbol | Common Name | Structure |
|--------|-------------|-----------|
| *Saturated Fatty Acids* | | |
| 12:0 | Lauric Acid | $CH_3(CH_2)_{10}COOH$ |
| 14:0 | Myristic Acid | $CH_3(CH_2)_{12}COOH$ |
| 16:0 | Palmitic Acid | $CH_3(CH_2)_{14}COOH$ |
| 18:0 | Stearic Acid | $CH_3(CH_2)_{16}COOH$ |

| Symbol | Common Name | Structure |
|--------|-------------|-----------|
| *Unsaturated Fatty Acids* | | |
| 16:1 | Palmitoleic Acid | $CH_3(CH_2)_5CH = CH(CH_2)_7COOH$ |
| 18:1 | Oleic Acid | $CH_3(CH_2)_7CH = CH(CH_2)_7COOH$ |
| 18:2 | Linoleic Acid | $CH_3(CH_2)_4(CH = CHCH_2)_2(CH_2)_6COOH$ |

Slide #6 – Saturated vs. unsaturated fatty acids. Saturated chains pack tightly and form more rigid, organized aggregates (i.e., membranes). Unsaturated chains bend (~30°) and pack in a less ordered way with greater potential for motion. In unsaturated fatty acids, the first double bond generally appears between _____ and _____. Triple bonds rarely occur in fatty acids.
*Notes*

Slide #7 – Triacylglycerols. Fats and oils in plants and animals consist of mixtures of triacylglycerols (triglycerides). Triacylglycerols are nonpolar, water insoluble substances that consist of fatty acid triesters of _____. They function as energy reservoirs in animals (Most _____ form of carbon in nature with tight packing). They are the most abundant class of lipids. Triacylglycerols are not components of _____. Simple triacylglycerols contain only one type of fatty acid.
*Notes*

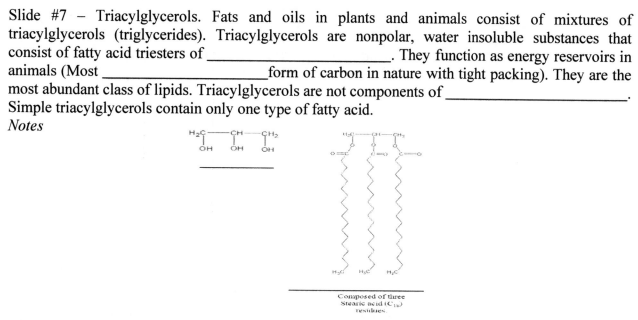

Composed of three
Stearic acid ($C_{18}$)
residues

Slide #8 – Mixed triacylglycerols contain two or three different types of fatty acid residues. They are named according to the placement on the glycerol moiety, and are much more common in nature than _____ triacylglycerols.
*Notes*

1-Palmitoleoyl-2-linoleoyl-3-stearoyl-
glycerol

Slide #9 – Polar bears produce all the water they need from metabolism of fat. Adult polar bears consume only fat (from seals they catch). By not consuming _____, they have no need to urinate or defecate and go for months without doing so, thus saving precious body water.
*Notes*

Slide #10 – _____ are major lipid components of biological membranes. They consist of sn-glycerol-3-phosphate esterified at its $C_1$ and $C_2$ positions to fatty acids. The phosphoryl group is covalently bound to group "X". Due to the fact that they have nonpolar aliphatic tails and polar phosphoryl head groups, they are said to be _____ molecules.
*Notes*

Slide #11 – Saturated $C_{16}$ and $C_{18}$ fatty acids usually occur at the _____ position. Unsaturated fatty acids ($C_{16}$-$C_{20}$) usually occur at the _____ position. Glycerphospholipids are named according to the identities of the fatty acids (and the identity of "X").
*Notes*

Slide #12 –
*Notes*

## Common Phospholipid Head Groups

| Formula of —— X | Name of Phospholipid |
|---|---|
| —— H | Phosphatidic Acid |
| —— $CH_2CH_2NH_3^+$ | Phosphatidylethanolamine |
| —— $CH_2CH_2N(CH_3)_3^+$ | Phosphatidylcholine |

Slide #13 – Naming.
*Notes*

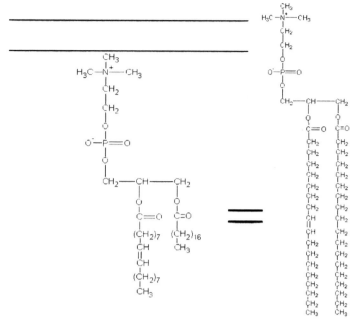

Slide #14 – What are the names of the following molecules?
*Notes*

1.)

2.)

_____

_____

_____

_____

Slide #15 – Ether glycerophospholipids contain an ether linkage at the _____ position of glycerol (instead of an ester linkage). Plasmalogens are ether glycerophospholipids in which the alkyl chain is _____. Platelet activating factor (PAF) is an ether glycerophospholipid in which the alkyl chain is _____.
*Notes*

_____        _____
Linkage                                Linkage

171

Slide #16 – Plasmalogens are glycerophospholipids in which the $C_1$ substituent of the glycerol moiety is bound to an _____ ether. The tails can vary in size. Choline plasmalogens are found in various tissues (i.e. _____ muscle).

*Notes*

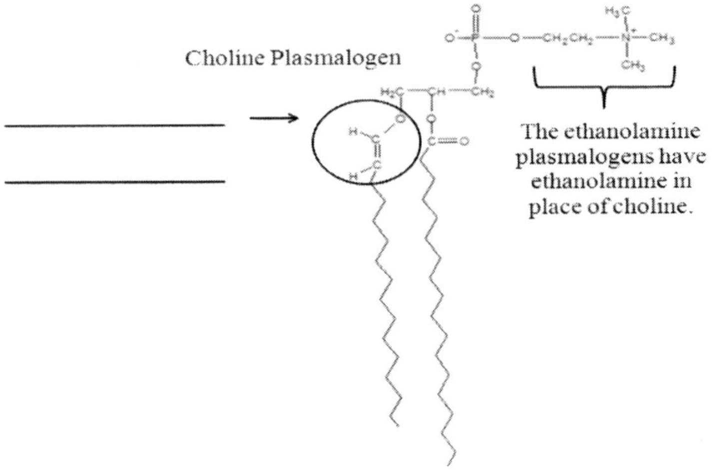

Choline Plasmalogen

The ethanolamine plasmalogens have ethanolamine in place of choline.

Slide #17 – Platelet Activating Factor (PAF) is an ether glycerophospholipid. PAF is a potent biochemical signal molecule (mediator of many leukocyte functions). Platelet Activating Factor has a short (_____) moiety at the $C_2$ position which allows for increased _____.

*Notes*

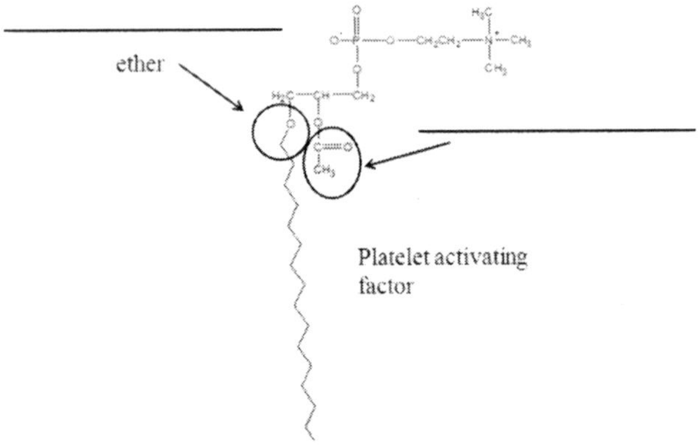

ether

Platelet activating factor

Slide #18 – Sphingolipids represent another class of lipids found frequently in biological membranes. Sphingolipids are derivatives of sphingosine ($C_{18}$ amino alcohols). The sphingosine moiety forms the backbone of these lipids rather than glycerol. A fatty acid joined to sphingosine in amide linkage forms a _____.

*Notes*

Slide #19 – Sphingolipids.
*Notes*

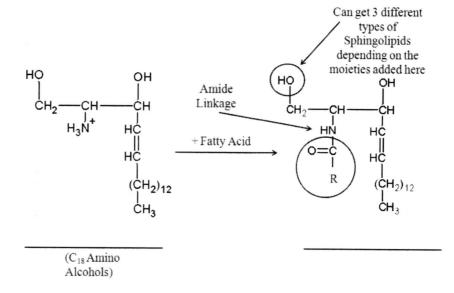

Slide #20 – _____ are the most common sphingolipid. They are ceramides with either a phosphocholine or a phosphoethanolamine. Therefore, they can be classified as a _____. They can be found in myelin sheaths that surrounds and insulates many nerve cell axons.
*Notes*

Slide #21 – _____ are the simpliest sphingoglycolipid. They are ceramides with one or more sugars in beta-glycosidic linkage at the 1-hydroxyl group. Galactocerebrosides (β-D-galactose head group) are prevalent in neuronal cell membranes of the brain, while glucocerebrosides (β-D-glucose residue) are found in cell membranes of various other tissues.
*Notes*

Slide #22 – _____ are the most complex group of the sphingoglycolipids. They are ceramide oligosaccharides, and must include among the sugar groups at least one _____ acid residue. Over 60 of these structures have been identified. They act as specific receptors for certain pituitary glycoprotein hormones that regulate physiologically important functions. Disorders involving the breakdown of this type of sphingoglycolipid are responsible for several hereditary diseases (i.e._____).
*Notes*

Slide #23 –
*Notes*

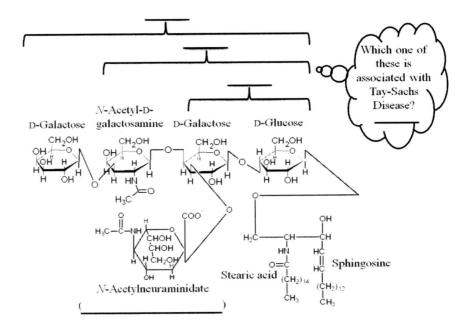

Slide #24 – Steroids are derivatives of a molecule containing four fused saturated rings. Cholesterol is a sterol lipid. It is the most common steroid in animals and precursor for all other steroids in animals. Examples include:

    a.) _____ provides control of carbohydrate, protein, and lipid metabolism.

    b.) _____ is the primary male sex steroid hormone.

    c.) _____ is the primary female sex steroid hormone.

    d.) _____ is a precursor of testosterone and estradiol.

*Notes*

Slide #25 – Lipid Metabolites Act As Biological Signals. Glycerophospholipids and sphingolipids play important roles as chemical signals in and on cells. These lipid signals act _____, either within the cell where they are made or on nearby cells (unlike steroid hormones). These signals typically initiate a cascade of reactions with multiple effects. The lifetimes of these signals are usually very _____. The creation and breakdown of lipid signals is carefully regulated and timed.
*Notes*

Slide #26 – Phospholipases A1 and A2 cleave fatty acids from a glycerophospholipid, producing _____. Phospholipases C and D hydrolyze on either side of the phosphate in the polar head group (This can produce various signaling molecules).
*Notes*

Phospholipase _____

Phospholipase _____

Phospholipase _____

Phospholipase _____

Slide #27 – _____ (S1P) is a phosphorylated sphingosine. S1P can exert a variety of intracellular effects. It can be exerted from the cell, where it can bind to membrane receptor proteins, either on adjacent cells, or the cell from which it was excreted. Excreted S1P binds to different receptor proteins and provokes many different cell and tissue effects.
*Notes*

Slide #28 – Lipid Aggregates. The first recorded experiments on the physical properties of lipids were made in 1774 by Benjamin Franklin. He investigated the well known action of oil calming waves (small waves). When oil is applied to the water, the hydrophilic head groups are immersed in the water while the hydrophobic tails extend into the air. In doing so, it disrupts water's surface tension, thereby calming the wave.

*Notes*

Slide #29 – _____ are formed by amphiphilic molecules (i.e. soaps and detergents). The molecular arrangement eliminates unfavorable interactions between water and the hydrophobic tails of the amphiphiles. Dilute aqueous solutions of amphiphiles do not form these. Thus, in order to get these you need to hit the Critical Micelle Concentration (_____), which is the concentration of amphiphile molecules required in order to form micelles.

*Notes*

Slide #30 – Micelles.
*Notes*

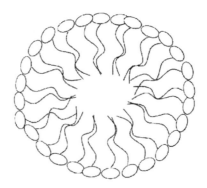

Slide #31 – _____ were first described in the 1960s by Alec Bangham. hey are formed from sonication (agitation with ultrasonic vibrations) of multilamellar vesicles (formed from suspending phospholipids in water) to yield unilamellar vesicles. Therefore, they have a phospholipid bilayer. Once formed they are very stable, and are generally stored at 4°C. Precise size can be controlled with an extruder.
*Notes*

Slide #32 – Liposomes.
*Notes*

Slide #33 – Micelles can accommodate _____ drugs in the internal hydrophobic core. Liposomes can accommodate both _____ as well as _____ chemotherapeutics either in the hydrophobic bilayer or the internal aqueous internal core respectively.
*Notes*

Slide #34 – Liposomes as Drug Delivery Vehicles in Cancer Therapy.
*Notes*

Slide #35 – Many early studies used liposomes as models to understand the cell bilayer. Using liposomes, it was determined early on that lipid bilayers are extremely impermeable to _____ and _____ substances. However, water (despite its polarity) is able to penetrate the bilayer due to its small size. With respect to diffusion, _____ diffusion (or flip-flop) is a rare event involving the transfer of lipid molecules across the bilayer to the other side. Another type of diffusion is _____ diffusion which is very common because lipids are highly mobile in the plane of the bilayer.
*Notes*

Slide #36 – The _____ temperature is the temperature at which phospholipids undergo a phase change (order-disorder transition). Above the transition temperature, the lipids are very mobile and "fluid-like". Below the transition temperature, the lipids are less mobile, and become a "gel-like" solid. These temperatures are unique to each phospholipid. The transition temperature of a lipid bilayer _____ with the chain length and degree of saturation.
*Notes*

Slide #37 – A look at what happens to the lipid bilayer at temperatures above and below the transition temperature.
*Notes*

Slide #38 – Biological membranes are composed of proteins associated with a lipid matrix. Specific proteins occur only in particular membranes. _____ (intrinsic) proteins are tightly bound to membranes via _____ forces, while _____ (extrinsic) proteins associate with the membrane by binding at its surface to lipid head groups and/or integral proteins through _____ interactions or hydrogen bonding.
*Notes*

Slide #39 – Biological Membranes.
*Notes*

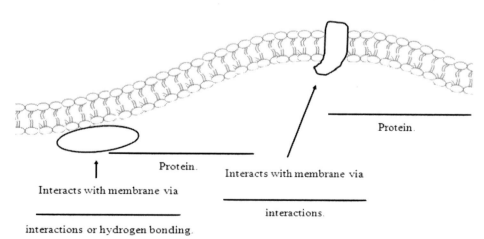

Slide #40 – _____ are compounds that cannot be synthesized in sufficient quantities by an organism. Therefore, they have to be obtained from the diet. They are classified by their biological and chemical activity, not their structure. They can be either water soluble or fat soluble. The Fat (lipid) soluble vitamins are _____, _____, _____, and _____. These are of interest in this chapter because they are soluble in lipids (hydrophobic).
*Notes*

Slide #41 – Vitamin _____ is also known as retinol. It is generated by enzymatic action in the liver from β-carotene. β-carotene is abundant in carrots (also occurs in various other vegetables). It serves as the site of the primary photochemical reaction in _____. Deficiency of this vitamin can lead to night blindness, and even total blindness.
*Notes*

Slide #42 – Vitamin _____ exists in many forms, and is abundantly present in milk. It is formed from _____ by the action of ultraviolet radiation from the sun. It leads to calcium uptake by the bones. Deficiency of this vitamin can lead to _____ (bones of growing children become soft). Adults exposed to normal amounts of sunlight usually do not require vitamin supplements.
*Notes*

Slide #43 – Vitamin _____, the most active of which is α-_____. It is an antioxidant. Therefore, it reacts with oxidizing agents before they can attack other biomolecules. The roles and importance of all of the various forms of this vitamin are presently unclear.
*Notes*

Slide #44 – Vitamin _____ comes from the Danish word _____. There are several forms of vitamin K that can be found within a single organism. The reason for this variation in poorly understood. It is an important factor in the blood-clotting process, which involves many steps and many proteins.
*Notes*

Slide #45 – Review for Exam #2.
*Notes*

1.) The rigidity of the peptide group in a protein can be attributed to the _____ _____ of the peptide bond.

2.) The _____ of a peptide are the atoms that participate in peptide bonds while ignoring side chain groups.

3.) The torsion angle about the Cα-N bond is known as _____ while the torsion angle about the Cα -C bond is known as _____

4.) The angles that Phi and Psi can adopt in a peptide can be depicted in a diagram called a _____ .

Slide #46 – Review for Exam #2.

5.) This Ramachandran diagram represents a protein that probably consists mostly of right handed helices, left handed helices, poly-L-proline II helices, or Beta sheets?

*Notes*

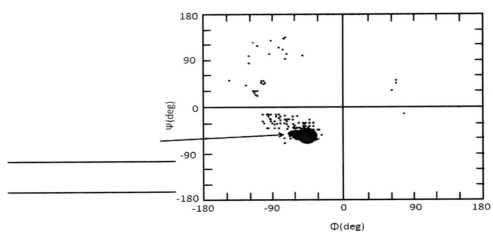

Slide #47 – Review for Exam #2.

6.) The following depiction represents a _____, in which Cα3 is usually a _____ residue and Cα2 is usually a _____ residue.

*Notes*

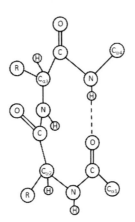

Slide #48 –

*Notes*

7.) In sugars, _____ and _____ membered rings are the most common due to greater stability.

8.) The name of the following molecule is _____ .

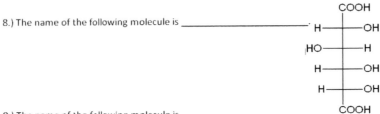

9.) The name of the following molecule is _____ .

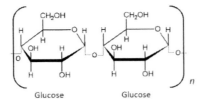

182

**Slide #49 –**
*Notes*

10.) The name of the following molecule is _____.

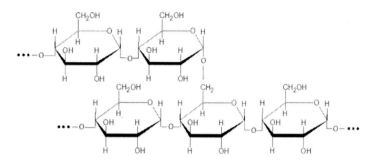

11.) The name of the following molecules from left to right is _____ and _____.

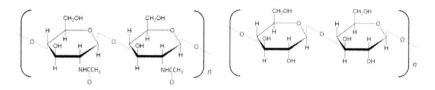

**Slide #50 –**
*Notes*

12.) What are fatty acids?

13.) What are triacylglycerols?

14.) What are phosphoacylglycerols?

15.) How does the composition of the bilayer affect its properties?

**Slide #51 –**
*Notes*

16.) What are sphingolipids?

17.) Name two lipid aggregates that are currently being used as drug delivery

18.) What are the fat soluble vitamins?

Slide #52 –
*Notes*

What are the names of the following molecules?

19.)

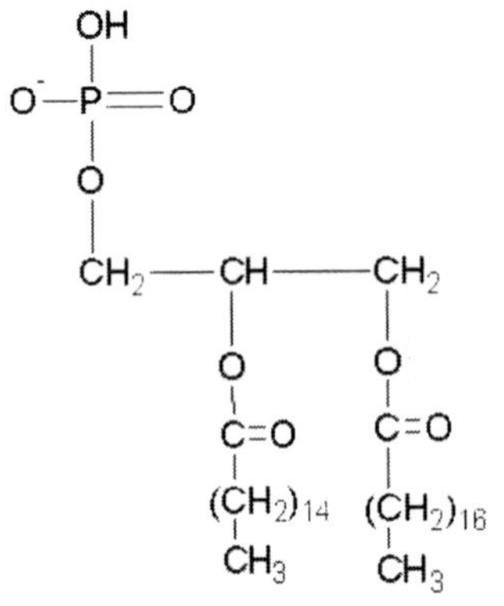

$O-\overset{OH}{\underset{O}{\overset{|}{P}}}=O$

$CH_2-CH-CH_2$

$\underset{C=O}{\overset{O}{|}}$  $\underset{C=O}{\overset{O}{|}}$

$(CH_2)_{14}$  $(CH_2)_{16}$

$CH_3$  $CH_3$

20.)

$NH_3^+$

$CH_2$

$CH_2$

$O$

$O-\overset{|}{\underset{O}{P}}=O$

$CH_2-CH-CH_2$

$\underset{C=O}{\overset{O}{|}}$  $\underset{C=O}{\overset{O}{|}}$

$(CH_2)_{16}$  $(CH_2)_{16}$

$CH_3$  $CH_3$

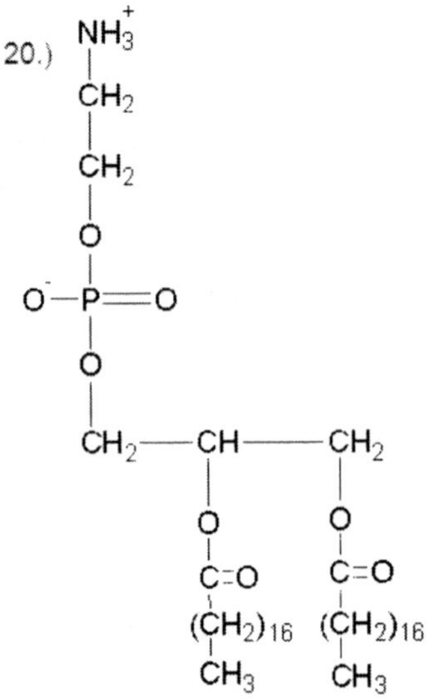

Slide #53 –
*Notes*

What is the name of the following molecule?

21.)

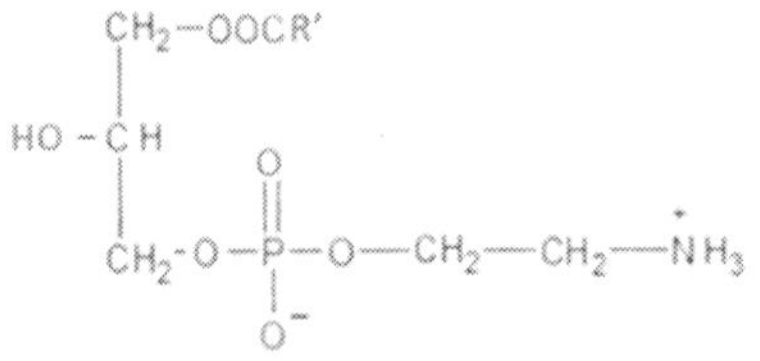

$CH_2-OOCR'$

$HO-CH$

$CH_2-O-\overset{O}{\underset{O^-}{\overset{||}{P}}}-O-CH_2-CH_2-\overset{+}{N}H_3$

Slide #54 – Recommended Problem Sets (from the textbook).
*Notes*

## *Chapter 7*
## *Additional Problem Sets (test format)*

1.) Which of the following is true regarding biological membranes?
   A.) They contain lipids and proteins.
   B.) They contain a small amount of carbohydrates.
   C.) They regulate the composition of the intracellular medium by controlling the flow of nutrients, waste products, ions etc. in and out of the cell.
   D.) They contain embedded pumps and gates in order to transport substances across the membrane.
   E.) All of the above.

2.) Which of the following is true regarding lipids?
   A.) The term "lipids" comes from the Greek word "lipos" for fat.
   B.) They are soluble in organic solvents.
   C.) They are sparingly soluble (if at all) in water.
   D.) Fats and oils are considered lipids.
   E.) All of the above.

3.) Which of the following statements about fatty acids is <u>not</u> true?
   A.) They are carboxylic acids with long-chain hydrocarbon side groups.
   B.) The hydrocarbon chains can be either saturated or unsaturated.
   C.) They are rarely found in free form.
   D.) They are mostly found with an even number of carbons.
   E.) All are true.

4.) Which one of the following statements about triacylglycerols is <u>not</u> true?
   A.) They consist of fatty acid triesters of glycerol.
   B.) They functions as energy reservoirs in animals.
   C.) They are the most abundant class of lipids.
   D.) They are the major lipid components of biological membranes.
   E.) All are true.

5.) Which of the following is true regarding glycerophospholipids?
   A.) They are the major lipid component of biological membranes.
   B.) They consist of *sn*-glycerol-3-phosphate esterified at its C1 and C2 positions to fatty acids.
   C.) The phosphoryl group is covalently bound to a head group "X", from which we can derive part of the name when naming glycerophospholipids (i.e. Phosphatidic acid, Phosphatidylethanolamine etc.).
   D.) Glycerophospholipids are amphiphilic molecules.
   E.) All of the above.

For Questions #6 - #10, consider the following structures:

A.)

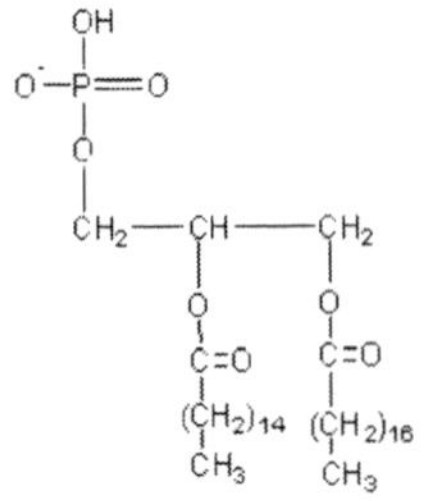

B.)

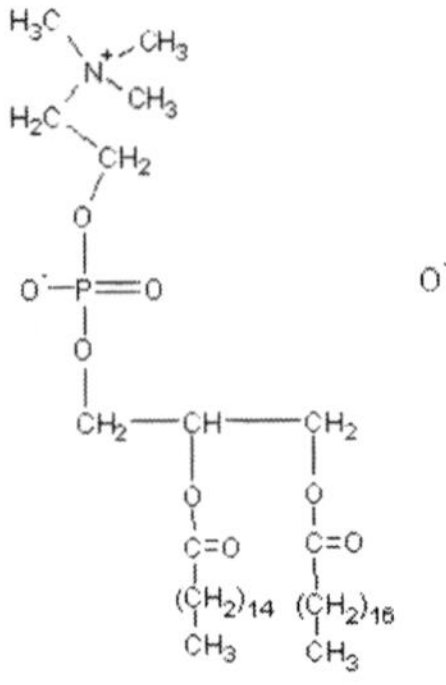

C.)

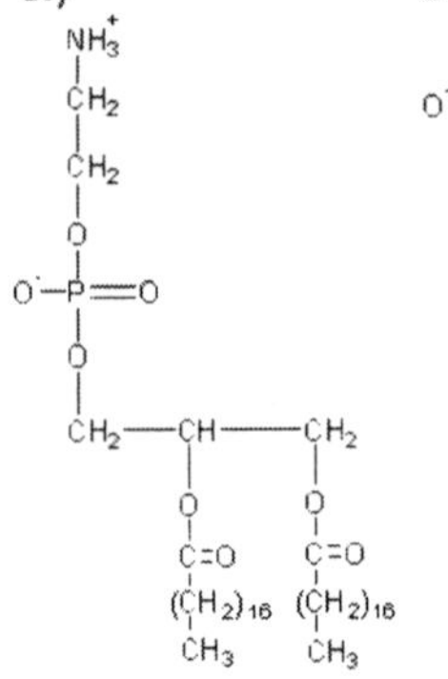

D.)

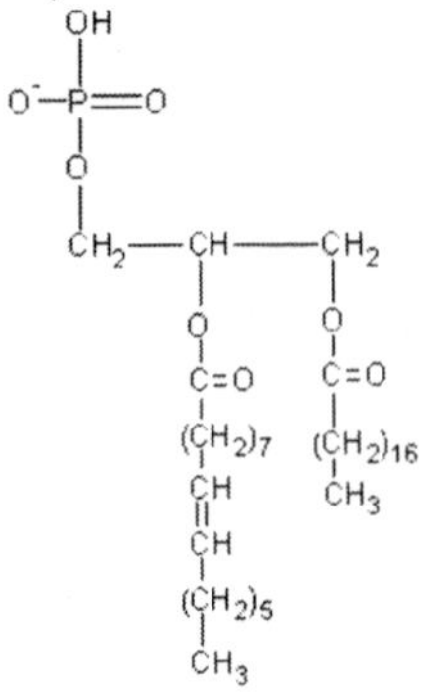

E.)

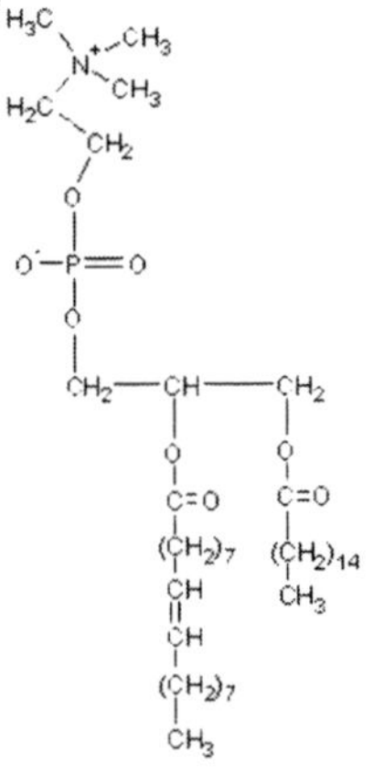

6.) Which of the above structures is 1-stearoyl-2-palmitoyl-3-phosphatidic acid?
  A.) A.
  B.) B.
  C.) C.
  D.) D.
  E.) E.

7.) Which of the above structures is 1-stearoyl-2-palmitoyl-3-phosphatidylcholine?
  A.) A.
  B.) B.
  C.) C.
  D.) D.
  E.) E.

8.) Which of the above structures is 1,2-distearoyl-3-phosphatidylethanolamine?
  A.) A.
  B.) B.
  C.) C.
  D.) D.
  E.) E.

9.) Which of the above structures is 1-stearoyl-2- palmitoleoyl-3-phosphatidic acid?
  A.) A.
  B.) B.
  C.) C.
  D.) D.
  E.) E.

10.) Which of the above structures is 1-palmitoyl-2-oleoyl-3- phosphatidylcholine?
    A.) A.
    B.) B.
    C.) C.
    D.) D.
    E.) E.

11.) As discussed in class, which of the following is true regarding ether glycerophospholipids?
    A.) Ether glycerophospholipids possess an ether linkage at the $C_1$ position of glycerol instead of an ester linkage (when compared to other unmodified glycerophospholipids).
    B.) Plasmalogens and platelet activating factor are both examples of ether glycerophospholipids.
    C.) Plasmalogens are ether glycerophospholipids in which the $C_1$ substituent of the glycerol moiety is bound to an unsaturated ether.
    D.) Platelet activating factor is an ether glycerophospholipid in which the alkyl chain is saturated.
    E.) All of the above.

12.) Which of the following statements about sphingolipids is <u>not</u> true?
    A.) Sphingolipids represent another class of lipids.
    B.) They are never found in biological membranes.
    C.) Sphingolipids are derivatives of sphingosine ($C_{18}$ amino alcohols).
    D.) A fatty acid joined to sphingosine in amide linkage forms a ceramide.
    E.) All are true.

13.) Which of the following is true regarding sphingomyelins?
    A.) They are the most common sphingolipid.
    B.) They can be composed of a ceramide moiety and a phosphocholine head group.
    C.) They can be composed of a ceramide moiety and a phosphoethanolamine head group.
    D.) They can be found in myelin sheaths that surrounds and insulates many nerve cell axons.
    E.) All of the above.

14.) Which of the following is true regarding cerebrosides?
    A.) They are the simplest sphingoglycolipid.
    B.) They are composed of a ceramide moiety and one or more sugars as part of the head group.
    C.) Galactocerebrosides contain a D-galactose head group.
    D.) Glucocerebrosides contain a D-glucose head group.
    E.) All of the above.

15.) Which of the following is true regarding gangliosides?
  A.) They are the most complex group of the sphingoglycolipids.
  B.) They can be considered ceramide oligosaccharides.
  C.) They have among the sugar groups at least one sialic acid residue.
  D.) Over 60 ganglioside structures have been identified.
  E.) All of the above.

For Questions #16 - #20, consider the following structures:

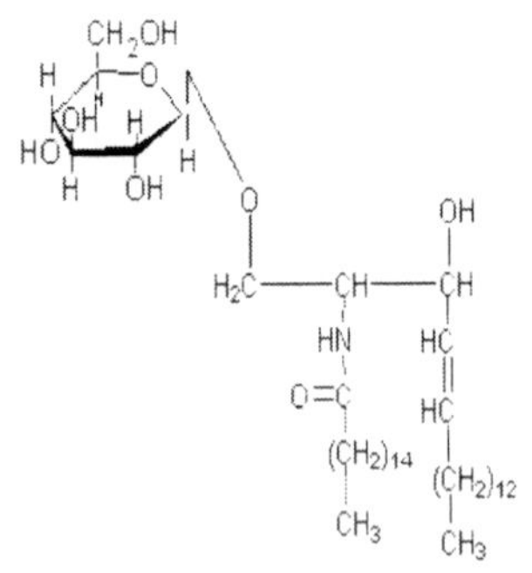

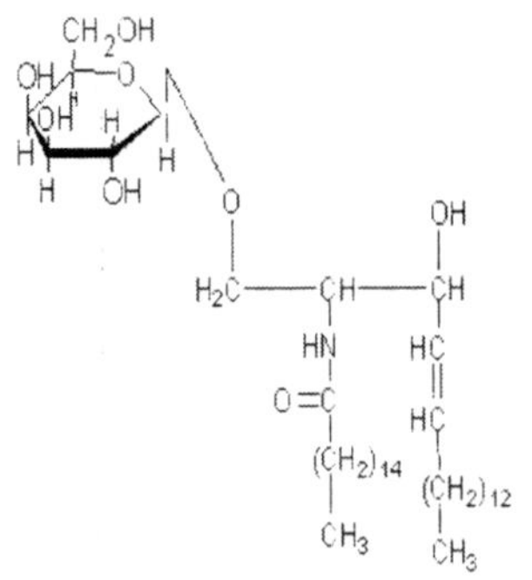

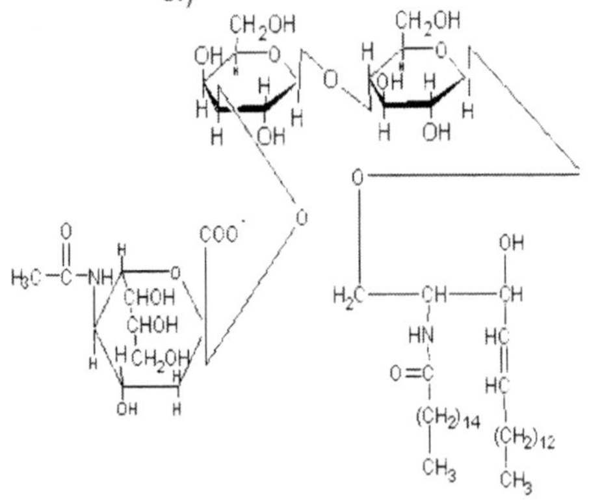

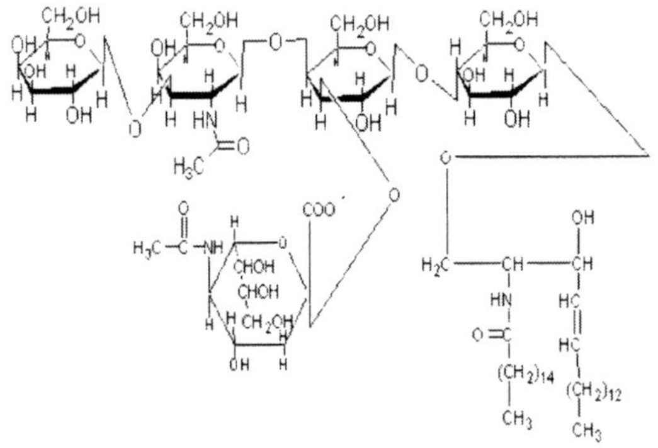

16.) Which of the above structures is ganglioside $G_{M1}$?
  A.) A.
  B.) B.
  C.) C.
  D.) D.
  E.) E.

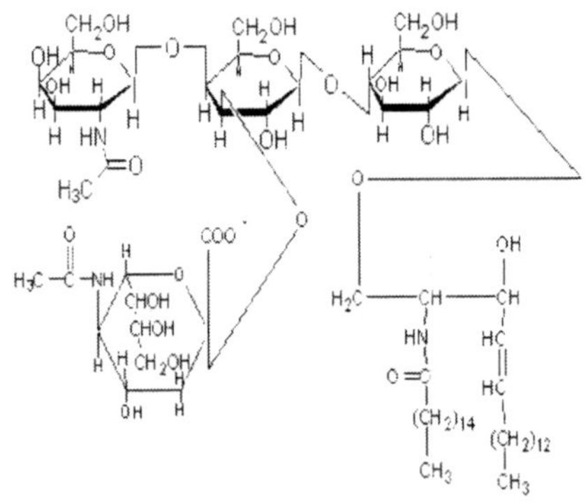

17.) Which of the above structures is a glucocerebroside (with a palmitate residue)?
  A.) A.
  B.) B.
  C.) C.
  D.) D.
  E.) E.

18.) Which of the above structures is ganglioside $G_{M2}$?
   A.) A.
   B.) B.
   C.) C.
   D.) D.
   E.) E.

19.) Which of the above structures is ganglioside $G_{M3}$?
   A.) A.
   B.) B.
   C.) C.
   D.) D.
   E.) E.

20.) Which of the above structures is a galactocerebroside (with a palmitate residue)?
   A.) A.
   B.) B.
   C.) C.
   D.) D.
   E.) E.

21.) _____ are the most complex group of sphingoglycolipids.
   A.) Sphingomyelins.
   B.) Gangliosides.
   C.) Cerebrosides.
   D.) Glucocerebrosides.
   E.) None of the above.

For Questions #22 - #25, consider the following phospholipid:

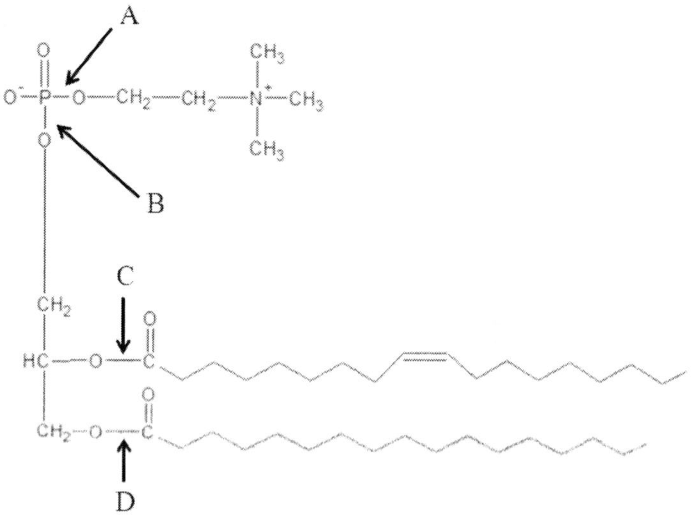

22.) Phospholipase _____ would hydrolyze the above phospholipid at point "A".
    A.) $A_1$.
    B.) $A_2$.
    C.) C.
    D.) D.
    E.) None of the above.

23.) Phospholipase _____ would hydrolyze the above phospholipid at point "B".
    A.) $A_1$.
    B.) $A_2$.
    C.) C.
    D.) D.
    E.) None of the above.

24.) Phospholipase _____ would hydrolyze the above phospholipid at point "C".
    A.) $A_1$.
    B.) $A_2$.
    C.) C.
    D.) D.
    E.) None of the above.

25.) Phospholipase _____ would hydrolyze the above phospholipid at point "D".
    A.) $A_1$.
    B.) $A_2$.
    C.) C.
    D.) D.
    E.) None of the above.

26.) As discussed in class, which one of the followings statements about biological signals attributed to lipid metabolites is <u>not</u> true?
    A.) They tend to act locally, either within the cell where they are made or on nearby cells.
    B.) These signals typically initiate a cascade of reactions with multiple effects.
    C.) The lifetimes of these signals are usually very long.
    D.) The creation and breakdown of lipid signals is carefully regulated and timed.
    E.) All are true.

27.) The _____ is the term traditionally used to identify the concentration of amphiphile molecules required in order to form micelles
    A.) Critical Concentration for Micelles (CCM).
    B.) Required Micelle Concentration (RMC).
    C.) Critical Micelle Concentration (CMC).
    D.) Essential Micelle Concentration (EMC).
    E.) None of the above.

28.) As discussed in class, _____ are drug delivery vehicles perfectly suited for carrying hydrophobic drugs in their internal hydrophobic core, while _____ are drug delivery vehicles that can accommodate both hydrophobic drugs in its bilayer, or hydrophilic drugs in its internal aqueous core.

    A.) Micelles; Liposomes.
    B.) Micelles; Micelles.
    C.) Liposomes; Micelles.
    D.) Liposomes; Liposomes.
    E.) None of the above.

29.) Which of the following is true regarding vitamins?

    A.) They are compounds that cannot be synthesized in sufficient quantities by an organism.
    B.) They have to be obtained from the diet.
    C.) They are generally classified by their biological and chemical activity.
    D.) Vitamins can be either water soluble or fat soluble.
    E.) All of the above.

30.) Which of the following is a fat soluble vitamin?

    A.) A.
    B.) D.
    C.) E.
    D.) K.
    E.) All of the above.

# Chapter 7
## *Answers to Additional Problem Sets (test format)*

1.) E
2.) E
3.) E
4.) D
5.) E
6.) A
7.) B
8.) C
9.) D
10.) E
11.) E
12.) B
13.) E
14.) E
15.) E
16.) E
17.) A
18.) D
19.) C
20.) B
21.) B
22.) D
23.) C
24.) B
25.) A
26.) C
27.) C
28.) A
29.) E
30.) E

# Chapter 8

## *Introduction to Enzymes*

*Chapter 8 Summary*:

*Enzymes.* Nearly all biochemical reactions that comprise life are mediated by biological catalysts known as enzymes. Enzymes differ from ordinary chemical catalysts. For example, reaction rates involving enzymes are several orders of magnitude greater than corresponding chemical catalyzed reactions. Also, enzymatically catalyzed reactions occur at relatively mild conditions (<100 °C, atmospheric conditions, and nearly neutral pH) compared to chemical catalysis which often requires elevated temperatures, high pressures and extreme of pH. In addition, enzymes have a greater degree of specificity with respect to both reactants and products. Enzymes also have a remarkable capacity for regulation (allosteric control, covalent modification, and variation of the amounts of enzymes synthesized). In chemical catalysis, most of the regulation has to do with the concentration of reactants and/or products. Most enzymes are globular proteins, and are highly specific (i.e. they can distinguish between stereoisomers of a given compound). Enzymes are not used up in the reaction. There are two models that have been developed to describe the formation of the enzyme/substrate complex. In the *Lock-and-Key* model, they substrate binds to that portion of the enzyme with a complementary shape. In the *Induced Fit* model, binding of the substrate induces a change in the conformation of the enzyme that results in a complementary fit. X-ray studies indicate that the substrate-binding sites of most enzymes are largely preformed, but most exhibit some degree of induced-fit. Substrates bind enzymes through noncovalent interactions (electrostatic, hydrogen bonding, hydrophobic interactions etc.). Generally, the substrate-binding site consists of an indentation on the surface of an enzyme molecule that is complementary in shape to the substrate (geometric complementarity). In addition, amino acid residues that form the binding site on the enzyme are arranged to interact specifically with the substrate in an attractive manner (electronic complementarity). Molecules that differ in shape or functional group distribution from the substrate cannot form enzyme/substrate complexes that lead to the formation of products. Enzymes are highly specific both in binding chiral substrates and catalyzing their reactions. Stereospecificity arises due to the fact that enzymes by virtue of their inherent chirality (proteins that consist of L-amino acids) form asymmetric active sites. A good example of this is trypsin which hydrolyzes polypeptides composed of only L-amino acids, not D-amino acids. In any event, we can say that enzymes are absolutely stereospecific in the reactions they catalyze.

*Cofactors* are nonprotein substances that take part in enzymatic reactions. They are known as the enzyme's "chemical teeth". Cofactors may be metal ions (i.e. $Zn^{2+}$ in the case of Carboxypeptidase A, which are required for catalytic activity. They can also be organic molecules known as *coenzymes* (i.e. $NAD^+$ in yeast alcohol dehydrogenase (YADH) which catalyzes the oxidation of small primary and secondary alcohols to their corresponding aldehydes or ketones respectively). It is important to recognize that coenzymes are chemically changed by enzymatic reactions. Thus, to complete the catalytic cycle, the coenzyme must be returned to its original state. Regeneration of the coenzyme is generally accomplished by the enzymatic activity of a different enzyme (i.e. a different reaction). Many water soluble vitamins are coenzyme precursors. The two vitamins discussed in class are folic acid and nicotinamide, which are necessary components of the coenzymes tetrahydrofolate and nicotinamide respectively.

*Enzymatic Regulation.* An organism must be able to regulate the catalytic activity of its enzymes. For example, this is necessary to coordinate its numerous metabolic processes, respond to changes in the environment, as well as to grow and differentiate. Enzymes can do this in two major ways. They can control enzyme availability, and/or control of enzyme activity (i.e. via structural or conformational alterations). With respect to enzyme availability, the rate of enzyme synthesis and degradation is directly controlled by the cell. A good example of this involves E. coli. For example, E.coli grown in the absence of the disaccharide lactose lack the enzyme needed to metabolize this sugar. However, within minutes of exposure to lactose, these bacteria commence synthesizing the enzyme required to utilize this nutrient. Similarly, cells can directly control the degradation of unneeded enzymes. As far as control of enzyme activity, structural or conformational alterations can directly regulate the catalytic activity of enzymes. The rate of an enzymatically catalyzed reaction is directly proportional to the enzyme-substrate complex (ES). The ES varies not only depending on the concentrations of both the enzyme and substrate, but also the enzyme's substrate-binding affinity. The catalytic activity of an enzyme can therefore be controlled through the variation of its substrate-binding affinity. *Allosteric regulation* involves a change in the shape and activity of an enzyme that results from molecular binding with a regulatory substance. Homotropic effects involve ligand binding that alters the binding affinity for the same ligand. For example, the binding of aspartate to aspartate transcarbamoylase (ATCase) is an example of a homotropic effect. *Heterotropic effects* involve ligand binding that alters the binding affinity for different ligands. For example, the binding of either cytodine triphosphate (CTP) (inhibitory) or adenosine triphosphate (ATP) (activates) to ATCase is an example of a heterotropic effect. ATCase catalyzes the formation of N-carbamoylaspartate. N-carbamoylaspartate formation is the first step unique to the biosynthesis of pyrimidines (C and T). Both substrates in the reaction (carbamoyl phosphate and aspartate) are both homotropic effectors. When rapid nucleic acid biosynthesis has depleted a cell's CTP pool, this effector dissociates from ATCase, thereby deinhibiting the enzyme and increasing CTP synthesis. If CTP is in excess, it inhibits ATCase, thereby reducing the rate of CTP production. The X-ray structure of ATCase was determined by William Lipscomb. William Lipscomb was awarded the Noble Prize in Chemistry in 1976. It was determined that ATCase has catalytic subunits arranged as two sets of trimers. The catalytic subunits are complexed with 3 sets of regulatory dimmers. Each of the regulatory dimers joins two catalytic subunits in different trimers. Dissociated catalytic trimers (catalytic/regulatory subunit separation) retains their catalytic activity, and exhibits a noncooperative (hyberbolic) substrate saturation curve. In fact, it has a maximum catalytic rate higher than that of the intact enzyme. It is unaffected by the presence of either ATP or CTP. The regulatory subunits bind these effectors, but have no catalytic activity associated with them. The conclusion is that the regulatory subunits allosterically reduce the activity of the catalytic subunits in the intact enzyme.

*Zymogens.* Another way to control the catalytic activity of enzymes includes zymogens. Zymogens are inactive precursors of an enzyme. They are sometimes referred to as a proenzyme. Unlike Allosteric interactions in which control of the enzyme is through reversible changes in the structure of the enzyme, zymogens are irreversibly transformed into an active enzyme. The active form of the enzyme is generated by cleavage of covalent bonds within the zymogen. The amino acid chain that is released upon activation is called the activation peptide. Some zymogens are named by adding "ogen" to the end of the enzyme name, and some have the word "Pro" added in front of the enzyme name. Trypsinogen and chymotrypsinogen are two classical examples of zymogens. Trypsinogen and Chymotrypsinogen are generated in the pancreas. The pancreas is protected from their catalytic activity as they are in the inactive zymogen form. They are activated in the small intestine (where their digestive properties are needed). The enzyme enteropeptidase converts trypsinogen to trypsin. Trypsin then converts chymotrypsinogen to chymotrypsin. The activation of zymogens plays a crucial role in the complex process of blood clot formation. In the final, best-characterized step of clot formation, the soluble protein fibrinogen is converted to the insoluble protein fibrin. Fibrinogen is converted to fibrin as a result of the proteolytic enzyme thrombin. Thrombin itself is produced from a zymogen (prothrombin).

*Enzyme Nomenclature.* Enzyme's are commonly named by adding the suffix "ase" to the name of the enzyme's substrate or to a phrase describing its catalytic action. For example, urease catalysis the hydrolysis of urea, and alcohol dehydrogenase catalyzes the oxidation of alcohols to their corresponding aldehydes. The International Union of Biochemistry and Molecular Biology (IUBMB) have adopted a method to provide a systematic name (as opposed to a recommended name) for each enzyme. The types of reactions based on the major classification number are as follows:

| Classification | Type of Reaction |
| --- | --- |
| 1. Oxidoreductases | Oxidation-reduction reactions |
| 2. Transferases | Transfer of functional groups |
| 3. Hydrolases | Hydrolysis reactions |
| 4. Lyases | Group elimination to form double bonds |
| 5. Isomerases | Isomerization |
| 6. Ligases | Bond formation (along with ATP hydrolysis) |

*Chapter 8*
*Lecture Series*

Slide #1 – Introduction
*Notes*

Slide #2 – Nearly all biochemical reactions that comprise life are mediated by biological catalysts known as enzymes. Enzymes differ from ordinary chemical catalysts in the following ways:

a.) Reaction rates are _____ orders of magnitude greater than corresponding chemical catalyzed reactions.

b.) Enzymatically catalyzed reactions occur at relatively _____ conditions (i.e. less than 100 °C, atmospheric conditions, and nearly neutral pH) compared to chemical catalysis, which often requires elevated temperatures and pressures and extremes of pH.
*Notes*

Slide #3 – Enzymes (Continued).

c.) Enzymes have a greater degree of _____ with respect to both reactants and products (rarely have side products).

d.) Enzymes have a remarkable capacity for regulation (allosteric control, covalent modification, and variation of the amounts of enzymes synthesized). In chemical catalysis, most of the regulation has to do with the _____ of reactants and/or products. Most enzymes are globular proteins. They are highly specific (i.e. they can distinguish between _____of a given compound), and are not used up in the reaction.
*Notes*

Slide #4 – Two models have been developed to describe formation of the enzyme-substrate complex:

    a.) _____ model: substrate binds to that portion of the enzyme with a complementary shape

    b.) _____ model: binding of the substrate induces a change in the conformation of the enzyme that results in a complementary fit.

*Notes*

Slide #5 – A look at the two models of E-S formation.

*Notes*

Slide #6 – Formation of product. The enzyme substrate complex (ES) is formed. The product is then generated and released from the enzyme resulting in the regeneration of the enzyme (E) and the product (P).

*Notes*

Slide #7 – Substrate Specificity. Substrates bind enzymes through noncovalent interactions (i.e. _____, _____, and _____ etc.) Generally, the substrate-binding site consists of an indentation on the surface of an enzyme molecule that is complementary in shape to the substrate (_____ complementarity). In addition, amino acid residues that form the binding site on the enzyme are arranged to interact specifically with the substrate in an attractive manner (electronic complementarity).

*Notes*

Slide #8 – Molecules that differ in shape or functional group distribution from the substrate cannot form enzyme-substrate complexes that lead to the formation of products. We looked at two models. In the _____ model, the substrate-binding site on the enzyme may exist in the absence of bound substrate. However, in the _____ model, the substrate-binding site does not form until bound to the substrate. X-ray studies indicate that the substrate-binding sites of most enzymes are largely preformed, but most exhibit some degree of induced-fit.

*Notes*

Slide #9 – Enzymes are highly specific both in binding chiral substrates and catalyzing their reactions. Stereospecificity arises due to the fact that enzymes by virtue of their inherent _____ (proteins that consist only of L-amino acids) form asymmetric active sites. For example, _____ hydrolyzes polypeptides composed of only L-amino acids, not D-amino acids. Thus, enzymes are absolutely stereospecific in the reactions they catalyze.

*Notes*

Slide #10 – _____ Specificity involves the selectivity regarding the identities of the chemical groups on the substrate. It is often a more stringent requirement than the stereo-specificity. Enzymes vary considerably in their degree of geometric specificity (some specific for only one compound, and others catalyze the reactions of a small range of related compounds). Yeast alcohol dehydrogenase (_____) catalyzes the oxidation of small primary and secondary alcohols to the their corresponding aldehydes or ketones respectively. Some digestive enzymes such as the enzymes discussed earlier in the semester (_____) are so permissive in their ranges of acceptable substrates, that their geometric specificities are more accurately described as preferences.

*Notes*

Slide #11 – _____ are nonprotein substances that take part in enzymatic reactions. They are known as the enzyme's "chemical teeth". They may be metal ions (i.e. $Zn^{2+}$ in the case of Carboxypeptidase A), which are required for catalytic activity. They can also be organic molecules known as _____ (which are transiently associated with the enzyme, such as $NAD^+$ in YADH). Other cofactors are essentially permanently associated with a protein are known as _____ groups (i.e. the heme group in hemoglobin).

*Notes*

Slide #12 – _____ is a coenzyme in reactions involving YADH. Coenzymes are chemically changed by enzymatic reactions. To complete the catalytic cycle, the coenzyme must be returned to its original state (usually different reaction).

*Notes*

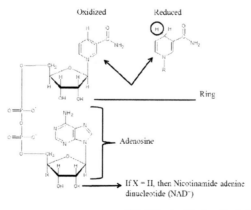

199

Slide #13 – Many vitamins (water soluble only) are coenzyme precursors. Many organisms are unable to synthesize certain portions of coenzymes and therefore these substances must be obtained from the diet (vitamins). Which two are we going to focus on?
*Notes*

| Vitamin | Coenzyme |
| --- | --- |
| 1.) | |
| 2.) | |

Slide #14 –
*Notes*

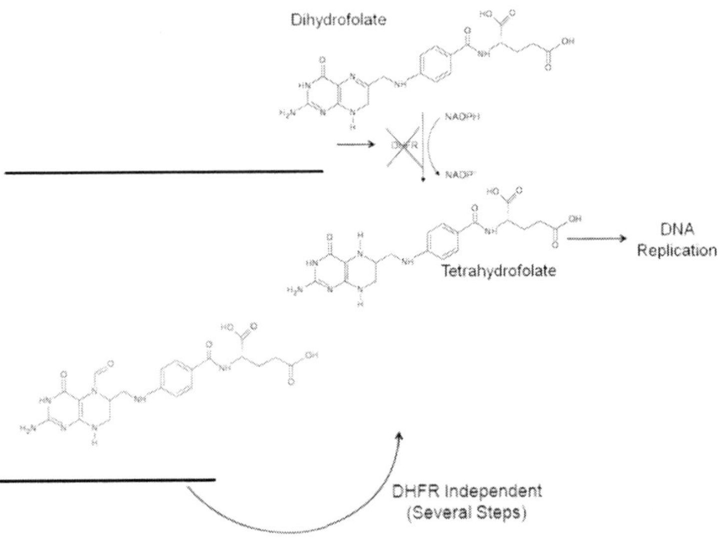

Dihydrofolate

NADPH

NADP

Tetrahydrofolate → DNA Replication

DHFR Independent (Several Steps)

Slide #15 – _____ was an endemic in the rural southern United States in the early twentieth century. It is characterized by diarrhea, dermatitis, and dementia. Results from vitamin B3 (nicotinic acid) deficiency. Most animals (including humans) can synthesize nicotinamide (component of $NAD^+$) from which amino acid? _____.
*Notes*

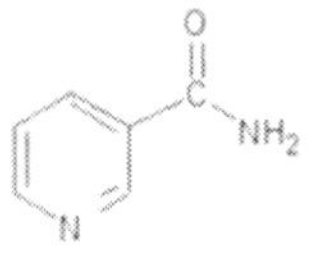

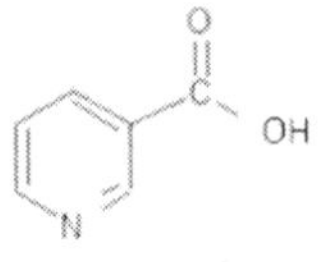

Slide #16 – An organism must be able to regulate the catalytic activity of its enzymes so that it can:

1.) _____.

2.) _____.

3.) _____.

*Notes*

Slide #17 – Regulation of the catalytic activity of enzymes can occur in two different ways. 1.) Control of enzyme availability (the amount of a given enzyme depends on both the rate of synthesis, as well as the rate of degradation). 2.) Control of enzyme activity (structural or _____ alterations can directly regulate the catalytic activity of enzymes).
*Notes*

Slide #18 – The rate of enzyme synthesis and degradation is directly controlled by the cell. For example, E.coli grown in the absence of the disaccharide _____ lack the enzyme needed to metabolize this sugar. However, within in minutes of exposure to this sugar, these bacteria commence synthesizing the enzyme required to utilize this nutrient. Similarly, cells can directly control the degradation of unneeded enzymes.
*Notes*

Slide #19 – Structural or conformational alterations can directly regulate the catalytic activity of enzymes. The rate of an enzymatically catalyzed reaction is directly proportional to the _____-_____ complex (ES). The ES varies not only depending on the concentrations of both the enzyme and substrate, but also the enzyme's substrate-binding affinity. The catalytic activity of an enzyme can therefore be controlled through the variation of its substrate-binding affinity.

*Notes*

Slide #20 – _____ regulation involves a change in the shape and activity of an enzyme that results from molecular binding with a regulatory substance. _____ effects involves ligand binding that alters the binding affinity for the same ligand. An example of this is the binding of aspartate to aspartate transcarbamoylase (ATCase). _____ effects involves ligand binding that alters the binding affinity for different ligands. An example of this is the binding of either cytodine triphosphate (CTP) (inhibitory) or adenosine triphosphate (ATP) (activates) to ATCase.

*Notes*

Slide #21 – Aspartate Transcarbamoylase (ATCase) catalyzes the formation of N-carbamoylaspartate, which is the first step unique to the biosynthesis of pyrimidines (C and T). Both substrates in the reaction are _____ effectors.

*Notes*

Slide #22 – _____ curve typical of allosteric reactions is observed with no allosteric effectors and for CTP inhibition. However, ATP activation results in a rate curve that is _____.

*Notes*

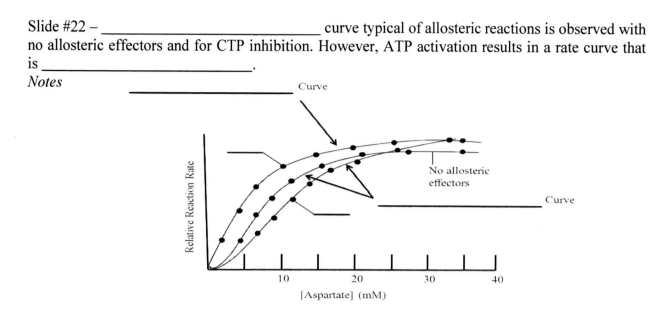

Slide #23 – When rapid nucleic acid biosynthesis has depleted a cell's _____ pool, this effector dissociates from ATCase, thereby deinhibiting the enzyme and increasing CTP synthesis. If CTP is in excess, it inhibits ATCase, thereby reducing the rate of CTP production.

*Notes*

Slide #24 – The X-ray structure of ATCase was determined by William Lipscomb. William Lipscomb was awarded the Noble Prize in Chemistry in 1976. It was determined that ATCase has _____ subunits arranged as two sets of trimers (c3) and 3 sets of _____ dimers (r2). Each of the regulatory dimers joins two catalytic subunits in different (c3) trimers.

*Notes*

Slide #25 – ATCase.
*Notes*

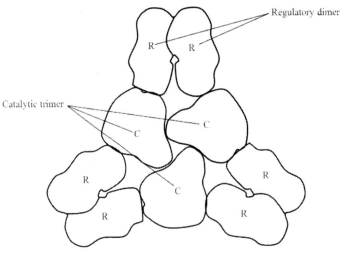

Regulatory subunit = "R" designation.

Catalytic subunit = "C" designation. Note that there is another set of trimer catalytic subunits behind this image and is therefore unseen in this depiction.

Slide #26 – Dissociated catalytic trimers (catalytic/regulatory subunit separation) retains their catalytic activity. It exhibits a _____ (hyberbolic) substrate saturation curve. It has a maximum catalytic rate _____ than that of the intact enzyme. It is unaffected by the presence of either ATP or CTP. The regulatory subunits bind these effectors, but have no catalytic activity associated with them. The conclusion is that the regulatory subunits _____ reduce the activity of the catalytic subunits in the intact enzyme.

*Notes*

Slide #27 – _____ Model for Allosteric Transitions.
*Notes*

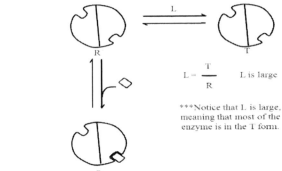

$$L = \frac{T}{R} \qquad \text{L is large}$$

***Notice that L is large, meaning that most of the enzyme is in the T form.

The equilibrium shifts here follows _____ Principle.

R = "_____" conformation (_____), which binds the substrate tightly.

T = "_____" or "_____" conformation (_____), which binds the substrate less tightly.

Slide #28 – _____ Model for Allosteric Transitions.
*Notes*

R = "_____" conformation (_____), which binds the substrate tightly.

T = "_____" or "_____" conformation (_____), which binds the substrate less tightly.

Slide #29 – _____ are another way to control the catalytic activity of enzymes. They are inactive precursors of an enzyme, sometimes referred to as a _____. Unlike allosteric interactions in which control of the enzyme is through reversible changes in the quaternary structure of the enzyme, these are _____ transformed into an active enzyme.
*Notes*

Slide #30 – The active form of the enzyme is generated by cleavage of covalent bonds. The amino acid chain that is released upon activation is called the _____ peptide. Some zymogens are named by adding "ogen" to the end of the enzyme name, and some have the word "Pro" added in front of the enzyme name. The two classical examples of zymogens discussed here are _____ and _____.
*Notes*

Slide #31 – Both of these zymogens are generated in the _____, which is protected from their catalytic activity as they are in the inactive zymogen form. They are activated in the _____ (where their digestive properties are needed). The enzyme enteropeptidase converts trypsinogen to trypsin. Trypsin then converts chymotrypsinogen to chymotrypsin.

*Notes*

Slide #32 – The activation of zymogens plays a crucial role in the complex process of blood clot formation. In the final, best-characterized step of clot formation, the soluble protein _____ is converted to the insoluble protein _____ as a result of the proteolytic enzyme thrombin. Thrombin itself is produced from a zymogen (prothrombin).

*Notes*

Slide #33 – Enzyme's are commonly named by adding the suffix "ase" to the name of the enzyme's substrate or to a phrase describing its catalytic action. For example, urease catalysis the hydrolysis of urea, and alcohol dehydrogenase catalyzes the oxidation of alcohols to their corresponding aldehydes. The International Union of Biochemistry and Molecular Biology (_____) have adopted a method to provide a systematic name (as opposed to a recommended name) for each enzyme.

*Notes*

Slide #34 – For the systematic name, it is the name of the enzyme's substrate, followed by a word ending "ase" which specifies the type of reaction. For example, carboxypeptidase A has the systematic name peptidyl-L-amino acid hydrolase and the classification number:

*Notes*

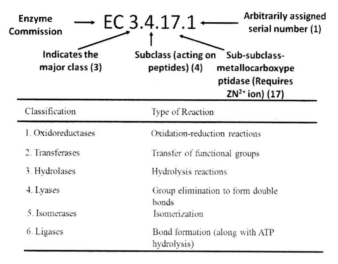

| Classification | Type of Reaction |
|---|---|
| 1. Oxidoreductases | Oxidation-reduction reactions |
| 2. Transferases | Transfer of functional groups |
| 3. Hydrolases | Hydrolysis reactions |
| 4. Lyases | Group elimination to form double bonds |
| 5. Isomerases | Isomerization |
| 6. Ligases | Bond formation (along with ATP hydrolysis) |

Slide #35 – Recommended Problem Sets (from the textbook).

*Notes*

## Chapter 8
### Additional Problem Sets (test format)

1.) When compared to ordinary chemical catalysts, which of the following are true regarding enzymes?
  A.) Enzymatically catalyzed reaction rates are several orders of magnitude greater than corresponding chemical catalyzed reactions.
  B.) Enzymatically catalyzed reactions occur at relatively mild conditions.
  C.) Enzymes have a greater degree of specificity.
  D.) Enzymes have a remarkable capacity for regulation, whereas in chemical catalysis most of the regulation has to do with the concentration of reactants and/or products.
  E.) All of the above.

2.) As discussed in class, which of the following is a model that describes the formation of the enzyme-substrate complex?
  A.) The Lock and Key Model.
  B.) The Induced-fit Model.
  C.) The Introduction Model.
  D.) Both A & B are correct.
  E.) None of the above.

3.) Which of the following is a true statement regarding the two models discussed in class describing the formation of the enzyme-substrate complex?
  A.) In the induced-fit model, a substrate binds to that portion of the enzyme with a preformed exact complementary shape.
  B.) In the induced-fit model, binding of the substrate induces a change in the conformation of the enzyme that results in a complementary fit.
  C.) The lock and key model discussed in class literally involves a real lock and key.
  D.) In the lock and key model, binding of the substrate induces a change in the conformation of the enzyme that results in a complementary fit.
  E.) None of the above.

4.) With respect to substrates binding to enzymes, which of the following are possible noncovalent forces that might be involved?
  A.) Electrostatic interactions.
  B.) Hydrogen bonding.
  C.) Hydrophobic interactions.
  D.) A, B, and C are all correct.
  E.) None of the above.

5.) Which of the following is true regarding substrate specificity?
  A.) Generally, the substrate binding site consists of an indentation on the surface of an enzyme molecule that is complementary in shape to the substrate.
  B.) Amino acid residues that form the binding site on the enzyme are arranged to interact specifically with the substrate in an attractive manner.
  C.) Molecules that differ in shape or functional group distribution from the substrate cannot form enzyme-substrate complexes.
  D.) Most enzymes are stereospecific in the reactions they catalyze.
  E.) All of the above.

6.) Which of the following is true regarding geometric specificity?
  A.) It is selectivity based on the identities of the chemical groups on the substrate.
  B.) It is often a more stringent requirement than the stereospecificity.
  C.) Enzymes vary considerably in their degree of geometric specificity.
  D.) Some digestive enzymes (i.e. Carboxypeptidase A) are permissive in their ranges of acceptable substrates.
  E.) All of the above.

7.) All of the following are properties of a coenzyme EXCEPT:
  A.) They are usually actively involved in the catalytic reaction of the enzyme.
  B.) They are chemically changed by enzymatic reactions.
  C.) They can serve as intermediate carriers of functional groups.
  D.) They are protein components.
  E.) All of the above are true.

8.) Which of the following is a known coenzyme in reactions involving yeast alcohol dehydrogenase (YADH).
  A.) $NIC^+$.
  B.) $NIX^+$.
  C.) $NAD^+$.
  D.) $NAP^+$.
  E.) None of the above.

9.) _____ is a "rescue drug" that can be given to patients receiving the chemotherapeutic agent methotrexate to alleviate the side-effects associated with the depletion of the coenzyme _____.
  A.) Leucovorin; tetrahydrofolate.
  B.) Tetrahydrofolate; leucovorin.
  C.) Tetrahydrofolate; folinic acid.
  D.) Nicotinamide; tetrahydrofolate.
  E.) None of the above.

For Questions #10 - #13, consider the following structure of NAD$^+$.

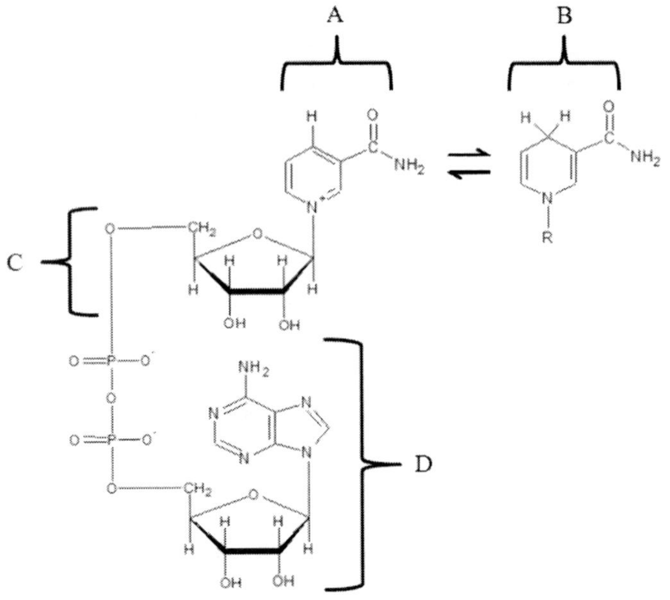

10.) The moiety labeled "A" in the above structure is _____.
   A.) D-Ribose.
   B.) Adenosine.
   C.) Nicotinamide (oxidized form).
   D.) Nicotinamide (reduced form).
   E.) None of the above.

11.) The moiety labeled "B" in the above structure is _____.
   A.) D-Ribose.
   B.) Adenosine.
   C.) Nicotinamide (oxidized form).
   D.) Nicotinamide (reduced form).
   E.) None of the above.

12.) The moiety labeled "C" in the above structure is _____.
   A.) D-Ribose.
   B.) Adenosine.
   C.) Nicotinamide (oxidized form).
   D.) Nicotinamide (reduced form).
   E.) None of the above.

13.) The moiety labeled "D" in the above structure is _____.
   A.) D-Ribose.
   B.) Adenosine.
   C.) Nicotinamide (oxidized form).
   D.) Nicotinamide (reduced form).
   E.) None of the above.

14.) Which of the following is true regarding pellagra?
   A.) It was an endemic in the rural southern United States in the early twentieth century.
   B.) It is caused by a chronic lack of vitamin $B_3$.
   C.) Symptoms can include diarrhea and dermatitis.
   D.) Symptoms can also include dementia.
   E.) All of the above.

15.) An organism must be able to regulate the catalytic activity of its enzymes so that it can
_____.
   A.) Coordinate its numerous metabolic processes.
   B.) Respond to changes in the environment.
   C.) Grow and differentiate.
   D.) A & B only.
   E.) All of the above.

For Questions #16 - #20, consider the following reaction catalyzed by ATCase (the first step in the pyrimidine biosynthetic pathway):

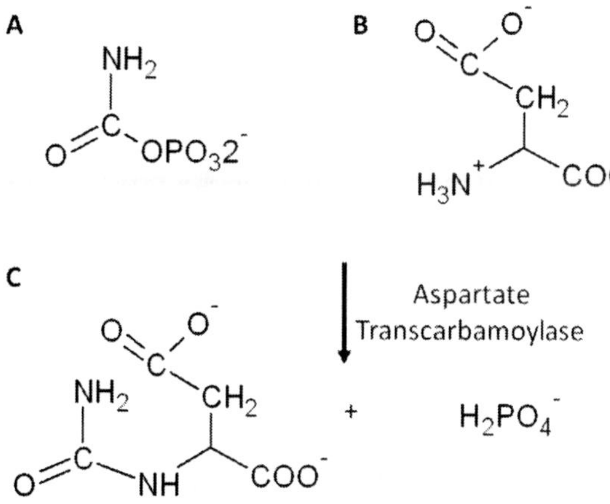

16.) In the above reaction, the structure labeled "A" is _____.
   A.) Aspartate.
   B.) Carbamoyl phosphate.
   C.) *N*-Carbamoylaspartate.
   D.) Both A & B.
   E.) None of the above.

17.) In the above reaction, the structure labeled "B" is _____.
   A.) Aspartate.
   B.) Carbamoyl phosphate.
   C.) *N*-Carbamoylaspartate.
   D.) Both A & B.
   E.) None of the above.

18.) In the above reaction, the structure labeled "C" is _____.
    A.) Aspartate.
    B.) Carbamoyl phosphate.
    C.) *N*-Carbamoylaspartate.
    D.) Both A & B.
    E.) None of the above.

19.) In the above reaction, _____ is a homotropic effector.
    A.) Aspartate.
    B.) Carbamoyl phosphate.
    C.) *N*-Carbamoylaspartate.
    D.) Both A & B.
    E.) None of the above.

20.) Which of the following is true regarding the enzymatic activity of ATCase in the above reaction?
    A.) Aspartate is a homotropic effector.
    B.) Carbamoyl phosphate is a homotropic effector.
    C.) It is heterotropically activated by ATP.
    D.) It is heterotropically inhibited by CTP.
    E.) All of the above.

21.) All of the following statements are true regarding the structure of ATCase EXCEPT:
    A.) It has catalytic subunits arranged as two sets of trimers.
    B.) The catalytic subunits are complexed with 2 sets of regulatory dimers.
    C.) Each of the regulatory dimers joins two catalytic subunits in different trimers.
    D.) It has a total of 6 individual catalytic subunits.
    E.) All of the above.

22.) The two principal models discussed in class for the behavior of allosteric enzymes are _____ and _____ models.
    A.) Concerted; sequester.
    B.) Concerted; sequential.
    C.) Concord; sequential.
    D.) Concord; sequester.
    E.) None of the above.

23.) Which of the following is true regarding zymogens?
    A.) They are another way to control the catalytic activity of enzymes.
    B.) They are inactive precursors of an enzyme.
    C.) They are sometimes referred to as a proenzyme.
    D.) They are irreversibly transformed into an active enzyme.
    E.) All of the above.

24.) As discussed in class, which of the following is true regarding trypsinogen and chymotrypsinogen?

    A.) They are generated in the pancreas.

    B.) The pancreas is protected from their catalytic activity as they are in the inactive zymogen form.

    C.) They are activated in the small intestine where their digestive properties are needed.

    D.) The enzyme enteropeptidase converts trypsinogen to trypsin, which can then convert chymotrypsinogen to chymotrypsin.

    E.) All of the above.

25.) According to the systemic nomenclature of enzymes as adopted by the IUBMB, and enzyme with a major classification number of 1 is involved in what type of reaction?

    A.) Oxidation-reduction reaction.

    B.) Transfer of functional groups.

    C.) Hydrolysis reaction.

    D.) Group elimination to form double bonds.

    E.) None of the above.

26.) According to the systemic nomenclature of enzymes as adopted by the IUBMB, and enzyme with a major classification number of 2 is involved in what type of reaction?

    A.) Oxidation-reduction reaction.

    B.) Transfer of functional groups.

    C.) Hydrolysis reaction.

    D.) Group elimination to form double bonds.

    E.) None of the above.

27.) According to the systemic nomenclature of enzymes as adopted by the IUBMB, and enzyme with a major classification number of 3 is involved in what type of reaction?

    A.) Oxidation-reduction reaction.

    B.) Transfer of functional groups.

    C.) Hydrolysis reaction.

    D.) Group elimination to form double bonds.

    E.) None of the above.

28.) According to the systemic nomenclature of enzymes as adopted by the IUBMB, and enzyme with a major classification number of 4 is involved in what type of reaction?

    A.) Oxidation-reduction reaction.

    B.) Transfer of functional groups.

    C.) Hydrolysis reaction.

    D.) Group elimination to form double bonds.

    E.) None of the above.

29.) According to the systemic nomenclature of enzymes as adopted by the IUBMB, and enzyme with a major classification number of 5 is involved in what type of reaction?
   A.) Oxidation-reduction reaction.
   B.) Transfer of functional groups.
   C.) Hydrolysis reaction.
   D.) Group elimination to form double bonds.
   E.) None of the above.

30.) According to the systemic nomenclature of enzymes as adopted by the IUBMB, and enzyme with a major classification number of 6 is involved in what type of reaction?
   A.) Oxidation-reduction reaction.
   B.) Transfer of functional groups.
   C.) Hydrolysis reaction.
   D.) Group elimination to form double bonds.
   E.) None of the above.

# Chapter 8
## *Answers to Additional Problem Sets (test format)*

1.) E
2.) D
3.) B
4.) D
5.) E
6.) E
7.) D
8.) C
9.) A
10.) C
11.) D
12.) A
13.) B
14.) E
15.) E
16.) B
17.) A
18.) C
19.) D
20.) A
21.) B (it's 3 sets of regulatory dimers, not 2).
22.) B
23.) E
24.) E
25.) A
26.) B
27.) C
28.) D
29.) E (it would be an isomerization reaction).
30.) E (It would be a ligase).

# Chapter 9

## *Rates of Enzymatic Reactions*

*Chapter 9 Summary*:
*Enzyme kinetics* is a subject that is of enormous practical importance in biochemistry because the maximum catalytic rate of an enzyme can be determined as well as binding affinities of both substrates and inhibitors. In addition, an enzyme's catalytic mechanism can be determined. Furthermore, this is an area of importance as most enzymes function as members of metabolic pathways. As mentioned earlier in the semester, enzymes are individually highly specific for particular reactions. They are responsible for diverse reactions such as hydrolysis, polymerization, functional group transfer, oxidation-reduction, dehydration, and isomerization reactions. Kinetic measurements of enzymatically catalyzed reactions are among the most powerful techniques for determining the catalytic mechanisms of enzymes. In biochemistry, Michaelis–Menten kinetics is the most commonly used model of enzyme kinetics. However, before we go any further, let us consider the following:

$$E + S \underset{k_{-1}}{\overset{k_1}{\rightleftharpoons}} ES \xrightarrow{k_2} E + P$$

Where E is the enzyme, S is the substrate, ES is the enzyme/substrate complex, and P is the product. Also, $k_1$ is the rate constant for the formation of ES, $k_{-1}$ is the rate constant for the reverse reaction, and $k_2$ is the rate constant for the formation of P from the ES. We can now set the rate of formation of the ES equal to the rate of the breakdown of the ES according to the steady state theory, and manipulate slightly.

$$k_1[E][S] = k_{-1}[ES] + k_2[ES]$$

$$[E] = [E]_T - [ES]$$

$$\frac{k_1([E]_T - [ES])[S]}{k_1[ES]} = \frac{k_{-1}[ES] + k_2[ES]}{k_1[ES]}$$

$$\frac{([E]_T - [ES])[S]}{[ES]} = \frac{k_{-1} + k_2}{k_1} = K_M$$

$$\frac{[E]_T[S] - [ES][S]}{[ES]} = K_M$$

$$[E]_T[S] - [ES][S] = K_M[ES]$$

$$[E]_T[S] = [ES](K_M + [S])$$

$$1/[ES] = (K_M + [S])/[E]_T[S]$$

$$[ES] = \frac{[E]_T[S]}{K_M + [S]}$$

216

Ok, now we can introduce a new equation here because in the initial stages of a reaction, there is very little product. Thus, we assume the initial rate ($V_o$) is dependant only on the [ES] breakdown to form the product (i.e. $V_o = k_2[ES]$), and substitute into the above equation.

$$V_o = k_2[ES]$$
$$V_{init} = k_2[ES] = \frac{k_2[E]_T[S]}{K_M + [S]}$$

**Eventually, $V_o = V_{max} = k_2[E]_T$**

$$V_{init} = \frac{V_{max}[S]}{K_M + [S]}$$

In order to make it a little easier to accurately determine $V_{max}$ and $K_M$, we can further manipulate the equation to get the equation of a straight line, which can be depicted graphically as a Lineweaver-Burk double-reciprocal plot.

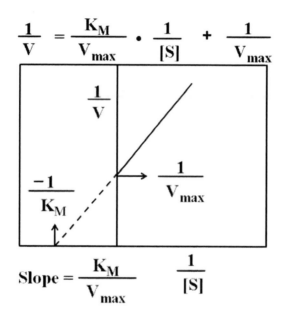

$$\frac{1}{V} = \frac{K_M}{V_{max}} \cdot \frac{1}{[S]} + \frac{1}{V_{max}}$$

In order to understand the significance of $V_{max}$, it is important to point out that it is related to the turnover number ($k_{cat}$), which is equal to $V_{max}/[ET]$. Thus, turnover numbers reported in the literature can be done without reporting enzyme concentrations. Turnover numbers tell you the number of moles of substrate that react to form product per mole enzyme per unit time. They are particularly dramatic illustration of the efficiency of an enzyme.

*Enzyme Inhibition.* A reversible inhibitor is a substance that binds to an enzyme to inhibit it, but can be released. The two kinds of inhibitors that we will discuss in class include both competitive and noncompetitive inhibitors. A competitive inhibitor binds to the active (catalytic) site and blocks the substrate from accessing it. A noncompetitive inhibitor binds to a site other than the active site and inhibits the enzyme by changing its conformation. In competitive inhibition $K_M$ changes (increases), and in noncompetitive inhibition $V_{max}$ changes (decreases). In the graphs below, [I] is the inhibitor concentration and $K_I$ is the dissociation constant for the enzyme/inhibitor complex.

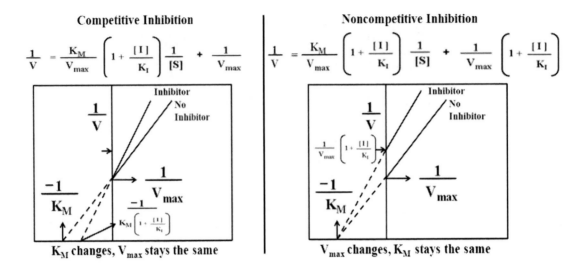

An irreversible inhibitor is a substance that causes inhibition that cannot be reversed. It usually involves formation or breaking of covalent bonds to or on the enzyme.

*Chapter 9*
*Lecture Series*

Slide #1 – Introduction
*Notes*

Slide #2 – Enzyme kinetics is a subject that is of enormous practical importance in biochemistry because _____ _____ of both substrates and inhibitors can be determined, maximum catalytic _____ of an enzyme can be determined, and an enzyme's catalytic _____ can be determined. In addition, most enzymes function as members of metabolic pathways.
*Notes*

Slide #3 – More on enzyme kinetics. Enzymes are individually highly specific for particular reactions. They are responsible for diverse reactions such as hydrolysis, polymerization, functional group transfer, oxidation-reduction, dehydration, and isomerization reactions. Kinetic _____ of enzymatically catalyzed reactions are among the most powerful techniques for determining the catalytic mechanisms of enzymes.
*Notes*

Slide #4 – E=Enzyme, S=Substrate, ES=Enzyme/Substrate Complex, P=Product. Initial velocity increases with increasing substrate concentration up to a point (when the substrate concentration is high enough to entirely convert the enzyme to the ES form).
*Notes*

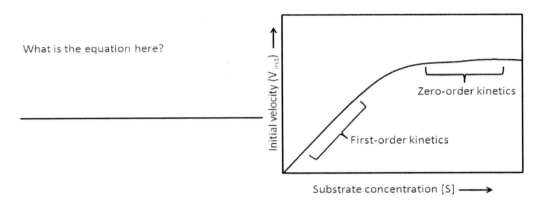

Slide #5 –
*Notes*

## Michaelis-Menten Kinetics.

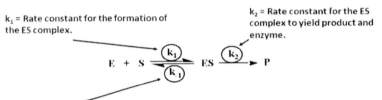

Therefore,

rate of formation of ES = $k_1$[E][S]

rate of breakdown of ES = $k_{-1}$[ES] + $k_2$[ES]

At steady state (Very little of the ES complex is present, and it turns over rapidly, but the concentration stays the same) rate of formation equals rate of breakdown. So,

Slide #6 –
*Notes*

## Michaelis-Menten Kinetics.

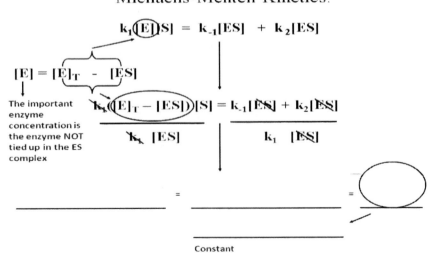

Slide #7 –
*Notes*

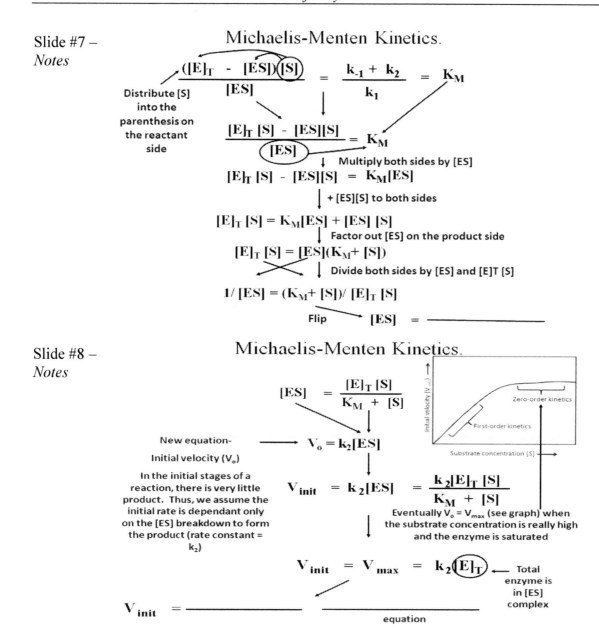

## Michaelis-Menten Kinetics.

$$\frac{([E]_T - [ES])[S]}{[ES]} = \frac{k_{-1} + k_2}{k_1} = K_M$$

Distribute [S] into the parenthesis on the reactant side

$$\frac{[E]_T [S] - [ES][S]}{[ES]} = K_M$$

Multiply both sides by [ES]

$$[E]_T [S] - [ES][S] = K_M[ES]$$

+ [ES][S] to both sides

$$[E]_T [S] = K_M[ES] + [ES] [S]$$

Factor out [ES] on the product side

$$[E]_T [S] = [ES](K_M + [S])$$

Divide both sides by [ES] and [E]T [S]

$$1/[ES] = (K_M + [S])/[E]_T [S]$$

Flip  $\quad [ES] = $ ———————

Slide #8 –
*Notes*

## Michaelis-Menten Kinetics.

$$[ES] = \frac{[E]_T [S]}{K_M + [S]}$$

New equation-

Initial velocity ($V_o$)

In the initial stages of a reaction, there is very little product. Thus, we assume the initial rate is dependant only on the [ES] breakdown to form the product (rate constant = $k_2$)

$$V_o = k_2[ES]$$

$$V_{init} = k_2[ES] = \frac{k_2[E]_T [S]}{K_M + [S]}$$

Eventually $V_o = V_{max}$ (see graph) when the substrate concentration is really high and the enzyme is saturated

$$V_{init} = V_{max} = k_2[E]_T$$ ← Total enzyme is in [ES] complex

$$V_{init} = \frac{\qquad}{\qquad}$$  equation

Slide #9 – Michaelis-Menten equation-Plot of initial velocity ($V_{init}$) vs. substrate concentration yields a _____ curve. $K_M$ = substrate concentration at ½ _____. $K_M$ tells us how tightly the substrate is bound (the greater the value, the _____ tightly bound). This curve makes it difficult to determine $V_{max}$. Therefore, it is difficult to determine $K_M$.
*Notes*

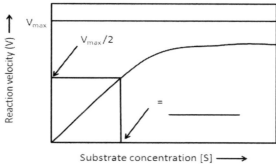

221

Slide #10 –
*Notes*

# Linearizing The Michaelis-Menten Equation.

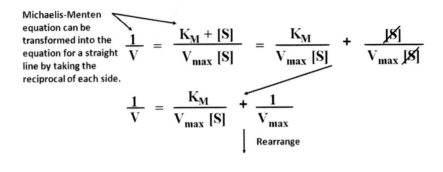

Michaelis-Menten equation can be transformed into the equation for a straight line by taking the reciprocal of each side.

$$\frac{1}{V} = \frac{K_M + [S]}{V_{max}\,[S]} = \frac{K_M}{V_{max}\,[S]} + \frac{[S]}{V_{max}\,[S]}$$

$$\frac{1}{V} = \frac{K_M}{V_{max}\,[S]} + \frac{1}{V_{max}}$$

Rearrange

1.) A plot of 1/V versus 1/[S] will give a straight line with slope of $K_M/V_{max}$ and y intercept of 1/Vmax

2.) This is known as a _____ double reciprocal plot

Slide #11 –
*Notes*

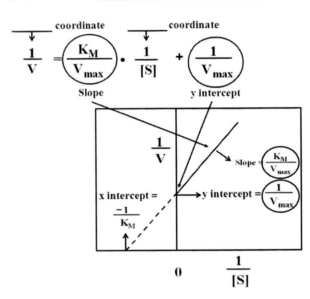

Slide #12 – Example 14.1-Calculate $K_M$ (mM) and $V_{max}$ (mM/sec) given the following information:
*Notes*

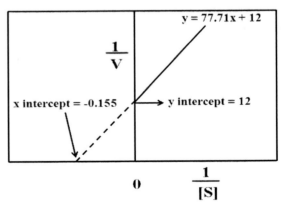

Slide #13 – Significance of $V_{max}$. $V_{max}$ is related to the turnover number, The turnover number is called _____, which is equal to (=) _____. Turnover numbers are reported in literature so that enzyme concentrations do not need be reported. Turnover numbers tell you the number of moles of substrate that react to form product per mole enzyme per unit time. Therefore, they are a particularly dramatic illustration of the efficiency of an enzyme.

*Notes*

Slide #14 – Enzyme Inhibition. Reversible inhibitor is a substance that binds to an enzyme to inhibit it, but can be released. The two reversible inhibitors that we are going to focus on include a _____ inhibitor which binds to the active (catalytic) site and blocks access to it so the substrate cannot bind, and a _____ inhibitor which binds to a site other than the active site (inhibits the enzyme by changing its conformation). An _____ inhibitor is a substance that causes inhibition that cannot be reversed. It usually involves formation or breaking of covalent bonds to or on the enzyme.

*Notes*

Slide #15 – Competitive vs. Noncompetitive. In both cases, the slope ($K_M/V_{max}$) in the Lineweaver-Burke Plot increases. A slope increase means a less efficient enzyme in the presence of either of both these inhibitors. The reason the slope changes can be explained by either an _____ in $K_M$ (_____ inhibition), or a _____ in $V_{max}$ (_____ inhibition). When $K_M$ changes, $V_{max}$ is unchanged and vice versa.

*Notes*

Slide #16 – Competitive Inhibition. $K_I$ = Disassociation constant, and [I] = inhibitor concentration. In a Lineweaver-Burk double reciprocal plot of 1/V versus 1/[S], the slope and the x intercept changes (_____ changes), but the y intercept does not change (_____).

*Notes*

**No Inhibition**

$$\frac{1}{V} = \frac{K_M}{V_{max}} \cdot \frac{1}{[S]} + \frac{1}{V_{max}}$$

**In the presence of a competitive inhibitor**

$$\frac{1}{V} = \frac{K_M}{V_{max}}\left(1 + \frac{[I]}{K_I}\right)\frac{1}{S} + \frac{1}{V_{max}}$$

Slide #17 – Lineweaver-Burke Plot for Competitive Inhibition.
*Notes*

Can adding more substrate potentially overcome this type of inhibition?

Yes        No

Slide #18 – Noncompetitive Inhibition. KI = Disassociation constant, and [I] = inhibitor concentration. In a Lineweaver-Burk double reciprocal plot of 1/V versus 1/[S], the slope and the y intercept changes (_____ changes), but because the inhibitor does not interfere with binding of substrate to the active site, the x intercept does not change (_____).
*Notes*

**No Inhibition**

$$\frac{1}{V} = \frac{K_M}{V_{max}} \cdot \frac{1}{[S]} + \frac{1}{V_{max}}$$

**In the presence of a noncompetitive inhibitor**

$$\frac{1}{V} = \frac{K_M}{V_{max}}\left(1 + \frac{[I]}{K_I}\right)\frac{1}{[S]} + \frac{1}{V_{max}}\left(1 + \frac{[I]}{K_I}\right)$$

Slide #19 – Lineweaver-Burke Plot for Noncompetitive Inhibition.
*Notes*

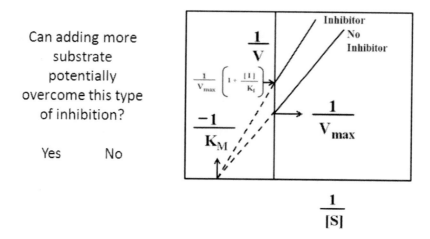

Can adding more substrate potentially overcome this type of inhibition?

Yes     No

Slide #20 – Recommended Problem Sets (from the textbook).
*Notes*

## Chapter 9
### Additional Problem Sets (test format)

1.) Enzyme kinetics is a subject that is of enormous practical importance in biochemistry because:
  A.) Binding affinities of both substrates and inhibitors can be determined.
  B.) Maximum catalytic rate of an enzyme can be determined.
  C.) An enzyme's catalytic mechanism can be determined.
  D.) Most enzymes function as members of metabolic pathways.
  E.) All of the above.

For Questions #2 - #7, consider the following:

$$E + S \underset{k_{-1}}{\overset{k_1}{\rightleftharpoons}} ES \xrightarrow{k_2} E + P$$

2.) The designation "E" in the above equation stands for _____.
  A.) Enzyme.
  B.) Substrate.
  C.) Enzyme/Substrate complex.
  D.) Product.
  E.) None of the above

3.) The designation "S" in the above equation stands for _____.
  A.) Enzyme.
  B.) Substrate.
  C.) Enzyme/Substrate complex.
  D.) Product.
  E.) None of the above

4.) The designation "ES" in the above equation stands for _____.
  A.) Enzyme.
  B.) Substrate.
  C.) Enzyme/Substrate complex.
  D.) Product.
  E.) None of the above

5.) The designation "P" in the above equation stands for _____.
   A.) Enzyme.
   B.) Substrate.
   C.) Enzyme/Substrate complex.
   D.) Product.
   E.) None of the above

6.) The designation "$k_1$" is the _____.
   A.) Enzyme.
   B.) Substrate.
   C.) Rate constant for the formation of ES.
   D.) Rate constant for the formation of P.
   E.) None of the above

7.) The designation "$k_2$" is the _____.
   A.) Enzyme.
   B.) Substrate.
   C.) Rate constant for the formation of ES.
   D.) Rate constant for the formation of P.
   E.) None of the above

8.) Which of the following is true regarding the Michaelis-Menton plot of initial velocity vs. substrate concentration?
   A.) The Michaelis-Menten equation plot yields a hyperbolic curve.
   B.) $K_M$ is equal to the substrate concentration at $\frac{1}{2} V_{max}$.
   C.) A hyperbolic curve makes it difficult to determine $V_{max}$.
   D.) A hyperbolic curve also makes it difficult to $K_M$.
   E.) All of the above.

For Questions #9 - #12, consider the following equation which can be expressed graphically to yield a Lineweaver-Burk double reciprocal plot:

$$\frac{1}{V} = \frac{K_M}{V_{max}} \cdot \frac{1}{[S]} + \frac{1}{V_{max}}$$

$$\downarrow \qquad \downarrow \qquad \downarrow \qquad \downarrow$$

$$\text{A} \qquad \text{B} \qquad \text{C} \qquad \text{D}$$

9.) In the above equation, "A" represents _____.
   A.) The x coordinate.
   B.) The y coordinate.
   C.) The slope.
   D.) The y intercept.
   E.) None of the above.

10.) In the above equation, "B" represents _____.
    A.) The x coordinate.
    B.) The y coordinate.
    C.) The slope.
    D.) The y intercept.
    E.) None of the above.

11.) In the above equation, "C" represents _____.
    A.) The x coordinate.
    B.) The y coordinate.
    C.) The slope.
    D.) The y intercept.
    E.) None of the above.

12.) In the above equation, "D" represents _____.
    A.) The x coordinate.
    B.) The y coordinate.
    C.) The slope.
    D.) The y intercept.
    E.) None of the above.

13.) Which of the following is true regarding turnover numbers in enzyme kinetics?
    A.) The turnover number is equal to $V_{max}/[ET]$.
    B.) Turnover numbers are reported in the literature so that enzyme concentrations need not be reported.
    C.) Turnover numbers tell you the number of moles of substrate that react to form product per mole enzyme per unit time.
    D.) Turnover numbers are a particularly dramatic illustration of the efficiency of an enzyme.
    E.) All of the above.

For Questions #14- #20, consider the following data:

| Substrate Concentration (mM) | Velocity (mM/min) |
| --- | --- |
| 4.0 | 2.8 |
| 1.0 | 2.0 |

14.) Given the data above, what would the x coordinates be in a graphical representation of a Lineweaver-Burk double reciprocal plot?

    A.) 0.36, 0.50.

    B.) 0.25, 1.0.

    C.) 0.91, 0.31

    D.) 4.0, 1.0.

    E.) None of the above.

15.) Given the data above, what would the y coordinates be in a graphical representation of a Lineweaver-Burk double reciprocal plot?

    A.) 0.36, 0.50.

    B.) 0.25, 1.0.

    C.) 0.91, 0.31

    D.) 4.0, 1.0.

    E.) None of the above.

16.) Given the data above, what would the slope be in a graphical representation of a Lineweaver-Burk double reciprocal plot?

    A.) 0.36.

    B.) 0.25.

    C.) 0.91.

    D.) 0.19.

    E.) None of the above.

17.) Given the data above, what would the y intercept be in a graphical representation of a Lineweaver-Burk double reciprocal plot?

    A.) 0.94.

    B.) 0.31.

    C.) 0.12.

    D.) 0.19.

    E.) None of the above.

18.) Given the data above, what would the x intercept be in a graphical representation of a Lineweaver-Burk double reciprocal plot?

    A.) -1.7.

    B.) -2.1.

    C.) -2.3.

    D.) -1.1.

    E.) None of the above.

19.) Given the data above, what is the value of $K_M$?
   A.) 0.34 mM.
   B.) 0.27 mM.
   C.) 0.13 mM.
   D.) 0.60 mM.
   E.) None of the above.

20.) Given the data above, what is the value of $V_{max}$?
   A.) 3.19 mM/min.
   B.) 4.27 mM/min.
   C.) 1.15 mM/min.
   D.) 2.58 mM/min.
   E.) None of the above.

For Questions #21- #27, consider the following data:

| Substrate Concentration (mM) | Velocity (mM/min) |
| --- | --- |
| 2.5 | 0.59 |
| 1.0 | 0.50 |

21.) Given the data above, what would the x coordinates be in a graphical representation of a Lineweaver-Burk double reciprocal plot?
   A.) 0.36, 0.50.
   B.) 0.25, 1.0.
   C.) 0.91, 0.31
   D.) 0.40, 1.0.
   E.) None of the above.

22.) Given the data above, what would the y coordinates be in a graphical representation of a Lineweaver-Burk double reciprocal plot?
   A.) 1.7, 2.0.
   B.) 2.6, 1.0.
   C.) 1.1, 0.31
   D.) 0.40, 1.0.
   E.) None of the above.

23.) Given the data above, what would the slope be in a graphical representation of a Lineweaver-Burk double reciprocal plot?
    A.) 1.7.
    B.) 0.12.
    C.) 0.50.
    D.) 0.31.
    E.) None of the above.

24.) Given the data above, what would the y intercept be in a graphical representation of a Lineweaver-Burk double reciprocal plot?
    A.) 1.1.
    B.) 0.12.
    C.) 1.5.
    D.) 0.31.
    E.) None of the above.

25.) Given the data above, what would the x intercept be in a graphical representation of a Lineweaver-Burk double reciprocal plot?
    A.) -2.9.
    B.) -1.6.
    C.) -2.5.
    D.) -0.31.
    E.) None of the above.

26.) Given the data above, what is the value of $K_M$?
    A.) 0.34 mM.
    B.) 0.27 mM.
    C.) 0.13 mM.
    D.) 0.60 mM.
    E.) None of the above.

27.) Given the data above, what is the value of $V_{max}$?
    A.) 0.13 mM/min.
    B.) 4.2 mM/min.
    C.) 0.67 mM/min.
    D.) 2.9 mM/min.
    E.) None of the above.

28.) Which of the following is true regarding competitive and noncompetitive inhibition?
    A.) In both cases the slope increases.
    B.) A slope increase means a less efficient enzyme.
    C.) The reason the slope changes can be explained by either and increase in $K_M$ (competitive inhibition), or a decrease in $V_{max}$ (noncompetitive inhibition).
    D.) A & B only.
    E.) All of the above.

For Questions #29- #30, consider the following:

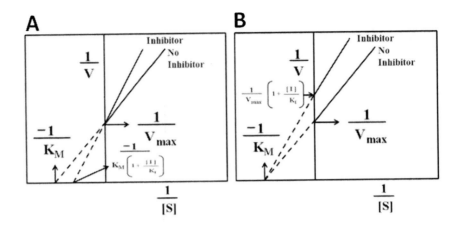

29.) The Lineweaver-Burk double reciprocal plot marked "A" above represents which kind of inhibition?

    A.) Competitive inhibition.

    B.) Noncompetitive inhibition.

    C.) Both competitive and noncompetitive inhibition.

    D.) There is no inhibition here.

    E.) None of the above.

30.) The Lineweaver-Burk double reciprocal plot marked "B" above represents which kind of inhibition?

    A.) Competitive inhibition.

    B.) Noncompetitive inhibition.

    C.) Both competitive and noncompetitive inhibition.

    D.) There is no inhibition here.

    E.) None of the above.

# Chapter 9
## *Answers to Additional Problem Sets (test format)*

1.) E
2.) A
3.) B
4.) C
5.) D
6.) C
7.) D
8.) E
9.) B
10.) C
11.) A
12.) D
13.) E
14.) B
15.) A
16.) D
17.) B
18.) A
19.) D
20.) A
21.) D
22.) A
23.) C
24.) C
25.) A
26.) A
27.) C
28.) E
29.) A
30.) B

# Chapter 10

## *Introduction to Metabolism*

*Chapter 10 Summary*:

*Metabolism* is the overall process through which living systems acquire and utilize the free energy they need to carry out their various functions. This is done by coupling exergonic reactions (reactions that release energy) of nutrient oxidation to the endergonic processes (processes that require energy) required to maintain the living state. Organisms show remarkable similarity in their major metabolic pathways, which would suggest that all life descended from a common ancestral form. However, living things also exhibit metabolic diversity. Organisms can be classified according to the major metabolic pathway in which they obtain carbon. For example, *autotrophs* are "self feeders" and are not dependant on an external source to produce complex organic compounds (i.e. they can use $CO_2$ as their sole source of carbon). *Heterotrophs* are organisms that require an organic form of carbon (i.e. glucose) in order to synthesize their other carbon compounds. Organisms can also be classified based on how they obtain their energy. *Phototrophs* (photosynthetic organisms) use light as a source of energy, while *chemotrophs* use organic compounds such as glucose to obtain their energy (typically through oxidation-reduction reactions). Chemotrophs obtain their free energy by oxidizing organic compounds (carbohydrates, lipids, and proteins). This free energy is most often coupled to endergonic reactions through the intermediate synthesis of "high energy" phosphate compounds (i.e. ATP-adenosine triphosphate). A series of metabolic reactions results in nutrients being oxidized to common intermediates (also called metabolites along with final products). These intermediates can then be used as precursors in the synthesis of other biological molecules. When describing metabolic reaction pathways, they are often divided into two categories, catabolism and anabolism. *Catabolism* involve degradative pathways and are usually energy yielding (constituents are broken down exergonically to salvage their components and/or to generate free energy). *Anabolism* on the other hand involve biosynthetic pathways in which biomolecules are synthesized from simpler components, and are usually energy requiring. We can use metabolic maps to portray the principal reactions of intermediary metabolism. When the major metabolic routes are known and functions are understood, the maps become easy to follow, despite their complexity. One interesting transformation of the metabolic map (simplified version) represents each intermediate as a black dot and each enzyme as a line. In this way, more than a thousand enzymes and substrates are represented by just two symbols. A dot connected to a single line must be a nutrient, a storage form, an end product, or an excretory product. A dot connected to just two lines is probably an intermediate in one pathway, while a dot connected to three represents an intermediate that usually has two metabolic fates. There are five principle characteristics associated with metabolic pathways. This includes the fact that metabolic pathways are irreversible, catabolic and anabolic pathways must differ, every metabolic pathway has a first committed step, all metabolic pathways are regulated, and metabolic pathways in eukaryotic cells occur in specific cellular locations (see following table that illustrates some important metabolic functions in eukaryotic cells and their specific cellular locations).

| Organelle | Function |
|---|---|
| Cytosol | Glycolysis (and many reactions involved in gluconeogenesis), fatty acid biosynthesis, pentose phosphate pathway. |
| Mitochondrion | Citric acid cycle, electron transport and oxidative phosphorylation, amino acid degradation, and fatty acid oxidation. |
| Nucleus | DNA replication and transcription (RNA processing ). |
| Rough Endoplasmic Reticulum | Protein synthesis. |

There many types of metabolic reactions to include redox reactions, group transfer reactions, eliminations, isomerizations, as well as rearrangement reactions. Regardless of the type of reaction, a very important aspect of metabolism involves *adenosine triphosphate (ATP)*, which is a "high-energy" intermediate that constitutes the most common cellular energy currency. It occurs in all known life-forms. The phosphoryl groups in ATP are linked via a phosphoester bond followed by two phosphoanhydride bonds. The rationale behind why ATP is a "high-energy" intermediate can be explained based on three major attributes associated with this compound. For example, the resonance stabilization energy of a phosphoanhydride bond is less than that of its hydrolysis products. Also, the destabilizing effect between the charged groups of a phosphoanhydride (electrostatic repulsion). Furthermore, the smaller solvation energy associated with the phosphoanhydride than the hydrolysis products all serve to explain why this compound is a "high-energy" intermediate.

## Chapter 10
### Lecture Series

Slide #1 – Introduction
*Notes*

Slide #2 – Metabolism is the overall process through which living systems acquire and utilize the free energy they need to carry out their various functions. This is done by coupling exergonic reactions (reactions that release energy) of nutrient oxidation to the endergonic processes (processes that require energy) required to maintain the living state. Organisms show remarkable similarity in their major metabolic pathways. This is would suggest that all life descended from a common ancestral form. However, living things also exhibit metabolic diversity.
*Notes*

Slide #3 – Organisms can be classified according to the major metabolic pathway in which they obtain _____. Autotrophs are organisms that are "self-feeders" (i.e. they can use $CO_2$ as their sole source of carbon). _____ are organisms that require an organic form of carbon (i.e. glucose) in order to synthesize their other carbon compounds. Organisms can also be classified based on how they obtain their energy. _____- (photosynthetic organisms) use light as a source of energy, while _____ use organic compounds such as glucose to obtain their energy (typically through oxidation-reduction reactions).
*Notes*

Slide #4 – _____ obtain their free energy by oxidizing organic compounds (carbohydrates, lipids, and proteins). This free energy is most often coupled to endergonic reactions through the intermediate synthesis of "high energy" phosphate compounds
(i.e. _____). A series of metabolic reactions results in nutrients being oxidized to common intermediates (also called metabolites along with final products). These intermediates can then be used as precursors in the synthesis of other biological molecules.

*Notes*

Slide #5 – The reaction pathways that comprise metabolism are often divided into two categories, _____ and _____.
_____ are degradative pathways, and are usually energy-yielding (constituents are broken down exergonically to salvage their components and/or to generate free energy). _____ involve biosynthetic pathways in which biomolecules are synthesized from simpler components, and are usually energy-requiring.

*Notes*

Slide #6 – Metabolic maps portray the principal reactions of intermediary metabolism. When the major metabolic routes are known and functions are understood, the maps become easy to follow, despite their complexity.

*Notes*

Slide #7 – One interesting transformation of the metabolic map (simplified) represents each intermediate as a black dot and each enzyme as a line. In this way, more than a thousand enzymes and substrates are represented by just two symbols. A dot connected to a single line must be a nutrient, a storage form, an end product, or an excretory product. A dot connected to just two lines is probably an intermediate in one pathway and has only one fate in metabolism. A dot connected to three represents an intermediate that has two metabolic fates.
*Notes*

Slide #8 – A look at a simplified metabolic map.
*Notes*

Slide #9 – There are five principle characteristics associated with metabolic pathways. Metabolic pathways are _____. Catabolic and anabolic pathways must _____. Every metabolic pathway has a first _____ step. All metabolic pathways are regulated. Metabolic pathways in eukaryotic cells occur in specific cellular locations.
*Notes*

Slide #10 – Metabolic pathways are _____. Highly exergonic reactions have a large negative free energy change. A large free energy change means that the reaction is irreversible. Irreversible means that these reactions go to_____. If such a reaction is part of a multistep pathway, it confers _____ to the pathway making the entire pathway irreversible.
*Notes*

Slide #11 – Catabolic and anabolic pathways must _____. If two metabolites are metabolically interconvertible, the pathway in going to from 1 to 2 must be different than going from 2 to 1. This is because if the conversion of 1 to 2 is exergonic, then the conversion from 2 to 1 must require energy (endergonic). The presence of two independent metabolic properties allows for independent control of the two processes.
*Notes*

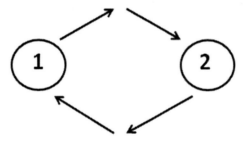

Slide #12 – Most reactions in a pathway function close to equilibrium. However, early in the metabolic pathway, there is generally an _____ exergonic reaction known as the first _____ step. This reaction "commits" the intermediate it produces to continue down the pathway.
*Notes*

Slide #13 – Metabolic pathways are regulated by supply and demand. A very efficient way to regulate a metabolic pathway is by regulating the _____ step. The first committed step (irreversible) functions too slowly for the reactant and products to equilibrate (would not be irreversible otherwise). Therefore, the first committed step if usually one of the rate-limiting steps. By controlling this step, the production of unnecessary _____ further along the pathway is avoided.

*Notes*

Slide #14 – Metabolic pathways can operate in different cellular locations due to the compartmentalization of eukaryotic cells (_____). A vital component of eukaryotic metabolism is their transport between these compartments (specific transport proteins in membranes), because they are often generated one place and utilized in another. ATP is generated in the _____, but much of it is used in the _____ of the cell.

*Notes*

Slide #15 – Eukaryotic Organelles.

*Notes*

| Organelle | Function |
| --- | --- |
| Cytosol | Glycolysis (and many reactions involved in gluconeogenesis), fatty acid biosynthesis, pentose phosphate pathway. |
| Mitochondrion | Citric acid cycle, electron transport and oxidative phosphorylation, amino acid degradation, and fatty acid oxidation. |
| Nucleus | DNA replication and transcription (RNA processing ). |
| Rough Endoplasmic Reticulum | Protein synthesis. |

Slide #16 –
*Notes*

# Types of Metabolic Reactions.

1.) _____ reactions.

2.) _____ reactions.

3.) Eliminations, isomerizations, and rearrangements.

4.) Reactions that make or break carbon-carbon bonds.

Slide #17 – Endergonic processes that maintain life are driven by the exergonic reactions of nutrient _____. Endergonic processes are driven by the "high-energy" intermediates that result from the _____ of nutrients. These intermediates are therefore a form of universal free energy "currency".
*Notes*

Slide #18 – Adenosine triphosphate (ATP) is a "high-energy" intermediate that constitutes the most common cellular energy currency. ATP occurs in all known life-forms. The ATP consists of an adenosine moiety with three phosphoryl groups. Can you label the ATP structure below?
*Notes*

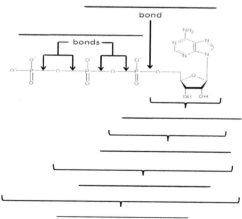

Slide #19 – Why is ATP such a high energy intermediate? _____ stabilization energy of a phosphoanhydride bond is less than that of its hydrolysis products. The destabilizing effect between the charged groups of a phosphoanhydride (electrostatic repulsion). Smaller _____ energy associated with the phosphoanhydride than hydrolysis products.

*Notes*

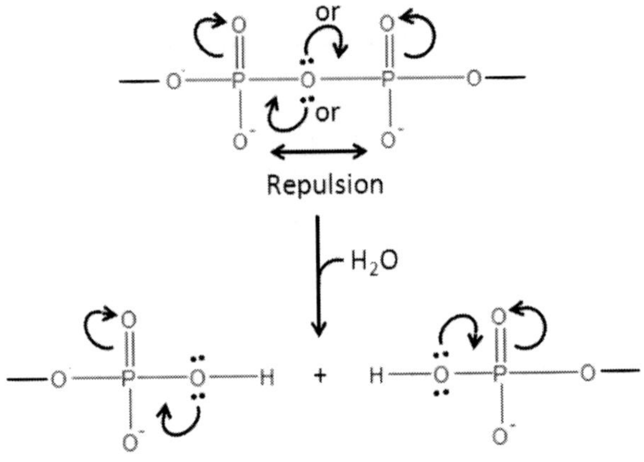

Slide #20 – Recommended Problem Sets (from the textbook).

*Notes*

## Chapter 10
### *Additional Problem Sets (test format)*

1.) Which of the following is true regarding metabolism?
  - A.) Metabolism is the overall process through which living systems acquire and utilize the free energy.
  - B.) This is done by coupling exergonic reactions of nutrient oxidation to endergonic processes.
  - C.) Organisms show remarkable similarity in their major metabolic pathways.
  - D.) Similarities observed in major metabolic pathways amongst organisms would suggest that all life descended from a common ancestral form.
  - E.) All of the above.

2.) Which of the following is true regarding metabolism classification?
  - A.) Organisms can be classified according to the major metabolic pathway in which they obtain carbon.
  - B.) Autotrophs are capable of synthesizing their own organic substances from inorganic compounds.
  - C.) Heterotrophs are organisms that require an organic form of carbon (i.e. glucose) in order to synthesize their other carbon compounds.
  - D.) Organisms can also be classified based on how they obtain their energy (i.e. phototrophs and chemotrophs).
  - E.) All of the above.

3.) Which of the following is true regarding metabolic reaction pathways?
  - A.) The reaction pathways that comprise metabolism are often divided into two categories, catabolism and anabolism.
  - B.) Catabolism involves degradative pathways.
  - C.) Anabolism involves biosynthetic pathways.
  - D.) Catabolic pathways are usually energy yielding, while anabolic pathways are usually energy requiring.
  - E.) All of the above.

For Questions #4 - #7, consider this depiction of a very small portion of a much larger "simplified" metabolic map:

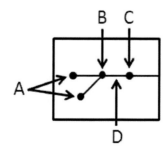

4.) In the above depiction, point "A" is probably a (n) _____.
   A.) Intermediate.
   B.) Enzyme.
   C.) Nutrient, a storage form, an end product, or an excretory product.
   D.) Intermediate that has two metabolic fates.
   E.) None of the above.

5.) In the above depiction, point "B" is probably a (n) _____.
   A.) Intermediate.
   B.) Enzyme.
   C.) Nutrient, a storage form, an end product, or an excretory product.
   D.) Intermediate that has two metabolic fates.
   E.) None of the above.

6.) In the above depiction, point "C" is probably a (n) _____.
   A.) Intermediate.
   B.) Enzyme.
   C.) Nutrient, a storage form, an end product, or an excretory product.
   D.) Intermediate that has two metabolic fates.
   E.) None of the above.

7.) In the above depiction, point "D" is probably a (n) _____.
   A.) Intermediate.
   B.) Enzyme.
   C.) Nutrient, a storage form, an end product, or an excretory product.
   D.) Intermediate that has two metabolic fates.
   E.) None of the above.

8.) Which of the following is NOT a characteristic of metabolic pathways?
   A.) Metabolic pathways are irreversible.
   B.) Catabolic and anabolic pathways must differ.
   C.) Every metabolic pathway has a first committed step.
   D.) All metabolic pathways are regulated.
   E.) All of the above are characteristic of metabolic pathways.

9.) Which of the following is true regarding the regulation of metabolic pathways?
   A.) Metabolic pathways are regulated by supply and demand.
   B.) A very efficient way to regulate a metabolic pathway is by regulating the rate-limiting step.
   C.) The first committed step is usually one of the rate-limiting steps.
   D.) By controlling this step, production of unnecessary metabolites further along the pathway is avoided.
   E.) All of the above.

10.) The citric acid cycle takes place in which organelle?
   A.) Cytosol.
   B.) Mitochondrion.
   C.) Nucleus.
   D.) Rough Endoplasmic Reticulum.
   E.) None of the above.

11.) Glycolysis takes place in which organelle?
   A.) Cytosol.
   B.) Mitochondrion.
   C.) Nucleus.
   D.) Rough Endoplasmic Reticulum.
   E.) None of the above.

12.) Electron transport and oxidative phosphorylation takes place in which organelle?
   A.) Cytosol.
   B.) Mitochondrion.
   C.) Nucleus.
   D.) Rough Endoplasmic Reticulum.
   E.) None of the above.

13.) DNA replication takes place in which organelle?
   A.) Cytosol.
   B.) Mitochondrion.
   C.) Nucleus.
   D.) Rough Endoplasmic Reticulum.
   E.) None of the above.

14.) Many of the reactions involved in gluconeogenesis takes place in which organelle?
   A.) Cytosol.
   B.) Mitochondrion.
   C.) Nucleus.
   D.) Rough Endoplasmic Reticulum.
   E.) None of the above.

15.) Fatty acid biosynthesis takes place in which organelle?
   A.) Cytosol.
   B.) Mitochondrion.
   C.) Nucleus.
   D.) Rough Endoplasmic Reticulum.
   E.) None of the above.

16.) DNA transcription takes place in which organelle?
    A.) Cytosol.
    B.) Mitochondrion.
    C.) Nucleus.
    D.) Rough Endoplasmic Reticulum.
    E.) None of the above.

17.) Fatty acid oxidation takes place in which organelle?
    A.) Cytosol.
    B.) Mitochondrion.
    C.) Nucleus.
    D.) Rough Endoplasmic Reticulum.
    E.) None of the above.

18.) The pentose phosphate pathway takes place in which organelle?
    A.) Cytosol.
    B.) Mitochondrion.
    C.) Nucleus.
    D.) Rough Endoplasmic Reticulum.
    E.) None of the above.

19.) Protein synthesis takes place in which organelle?
    A.) Cytosol.
    B.) Mitochondrion.
    C.) Nucleus.
    D.) Rough Endoplasmic Reticulum.
    E.) None of the above.

20.) Amino acid degradation takes place in which organelle?
    A.) Cytosol.
    B.) Mitochondrion.
    C.) Nucleus.
    D.) Rough Endoplasmic Reticulum.
    E.) None of the above.

21.) Which of the following are possible types of metabolic reactions?
    A.) Group transfer reactions.
    B.) Oxidation/reduction reactions.
    C.) Elimination reactions.
    D.) Isomerization reactions.
    E.) All of the above.

22.) Which of the following is true regarding ATP?
  A.) ATP stands for adenosine triphosphate.
  B.) It is a "high-energy" intermediate.
  C.) It is considered the most common cellular energy "currency".
  D.) It occurs in all known life forms.
  E.) All of the above.

23.) Which of the following is true regarding ATP?
  A.) ATP is continuously recycled in organisms.
  B.) It consists of an adenosine moiety with three phosphoryl groups.
  C.) The α phosphoryl group is linked via a phosphoester bond.
  D.) The other two phosphoryl groups (β and γ) are linked via phosphoanhydride bonds.
  E.) All of the above.

For Questions #24 - #29, consider the following structure:

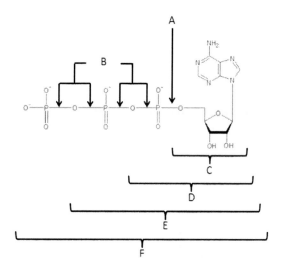

24.) In the above figure, "A" is (are) _____.
  A.) An adenosine moiety.
  B.) AMP.
  C.) A phosphoester bond.
  D.) Phosphoanhydride bonds.
  E.) None of the above.

25.) In the above figure, "B" is (are) _____.
  A.) An adenosine moiety.
  B.) AMP.
  C.) A phosphoester bond.
  D.) Phosphoanhydride bonds.
  E.) None of the above.

26.) In the above figure, "C" is (are) _____.
    A.) An adenosine moiety.
    B.) AMP.
    C.) ADP.
    D.) ATP.
    E.) None of the above.

27.) In the above figure, "D" is (are) _____.
    A.) An adenosine moiety.
    B.) AMP.
    C.) ADP.
    D.) ATP.
    E.) None of the above.

28.) In the above figure, "E" is (are) _____.
    A.) An adenosine moiety.
    B.) AMP.
    C.) ADP.
    D.) ATP.
    E.) None of the above.

29.) In the above figure, "F" is (are) _____.
    A.) An adenosine moiety.
    B.) AMP.
    C.) ADP.
    D.) ATP.
    E.) None of the above.

30.) Which of the following factors is responsible for the "high-energy" character of phosphoanhydride bonds such as those in ATP?
    A.) The resonance stabilization energy of a phosphoanhydride bond is less than that of its hydrolysis products.
    B.) The destabilizing effect between the charged groups of a phosphoanhydride (electrostatic repulsion).
    C.) The smaller solvation energy associated with the phosphoanhydride than hydrolysis products.
    D.) All of the above.
    E.) A.) and B.) only.

# Chapter 10
## *Answers to Additional Problem Sets (test format)*

1.) E
2.) E
3.) E
4.) C
5.) D
6.) A
7.) B
8.) E
9.) E
10.) B
11.) A
12.) B
13.) C
14.) A
15.) A
16.) C
17.) B
18.) A
19.) D
20.) B
21.) E
22.) E
23.) E
24.) C
25.) D
26.) A
27.) B
28.) C
29.) D
30.) D

CPSIA information can be obtained at www.ICGtesting.com
Printed in the USA
LVOW09s1447070915

453146LV00015B/345/P